ENGINEERING

AN ILLUSTRATED HISTORY FROM
ANCIENT CRAFT TO MODERN TECHNOLOGY

ENGINEERING

AN ILLUSTRATED HISTORY FROM
ANCIENT CRAFT TO MODERN TECHNOLOGY

Edited by Tom Jackson

Contributors: Tim Harris · Robert Snedden

SHELTER HARBOR PRESS

NEW YORK

Contents

Title Page

Left: In the 1st century CE, *the Romans perfected the art of dome-building with the Pantheon, an early example of concrete construction.*

Right: A scale model of an F-18 Super Hornet fighter is suspended inside a wind tunnel to test the way air flows around it at high speed.

Introduction

ENGINEERS ARE OFTEN OVERLOOKED IN THE STORIES OF GREAT EVENTS. THE SCIENTISTS WHO PUSH BACK THE BOUNDARIES OF KNOWLEDGE ARE RIGHTLY APPLAUDED, BUT IT IS THE ENGINEERS WHO TURN THAT KNOWLEDGE INTO SOMETHING THAT CHANGES THE WORLD. ENGINEERING IS THE APPLICATION OF SCIENCE, AND IT HAS BEEN IMPROVING OUR LIVES SINCE THE DAWN OF CIVILIZATION.

The thoughts and deeds of great achievers always make great stories, and here we have one hundred all together. Each story relates a ponderable, a weighty problem that became an invention and changed our homes, our cities, our farms, and our lives.

The word "engineer" means many things to many people. Some of us may think an engineer is the person who comes to repair our stuff when it goes wrong, but there is a lot more to engineering than that. In fact, the history of engineering is also the history of human civilization. Every civilization developed new technology, a particular set of tools, machines, and construction techniques. And each civilization—the Babylonians, ancient Egyptians, the Incas, and the Romans—all rose and fell thanks to their technology. Developing technology is the job of engineers. Engineers take what is known about the world and apply it in imaginative ways to solve problems. That might involve inventing a new machine or improving on an old process, and as scientists reveal the way the world works in ever more detail, so engineers can make better technology.

AN EARLY START

Technology is always being renewed and improved upon. Look around right now and you will see the fruits of centuries of engineering. In fact, engineering is older than humanity itself. Our distant relatives were making crude stone tools more than three million years ago—a technology that helped them survive.

BUILDING UP

The pace of technological advances was very slow back then—it took many hundreds of thousands of years for improved tools and new ways of doing things to appear, but we've been making up for the slow start ever since! Ancient civilizations were responsible for many of the greatest contributions to engineering. Ancient engineers invented boats; tamed fire; developed pottery, bricks, and even concrete; and learned to refine and use metals, such as copper and iron. And let's not forget the wheel, which appeared at least 5,000 years ago.

Engineers still rely on these breakthroughs even today: An engine tames the fire of burning fuel to create motion; a computer's spinning hard disk is just as much a wheel as the cogs in a clock or the millstones that grind wheat into flour for bread; and concrete and steel (a strengthened form of iron) have become the most commonly used materials in our constructions. They make it possible to build the widest bridges, tallest skyscrapers, and immense dams that hold back the world's largest rivers.

REVOLUTIONS

The pace of change in modern engineering is hard to fathom. It sometimes appears there is a perpetual revolution, with one technology always being superseded by another. We can trace this rate of change back to an original revolution.

It might not look like much, but this 2-million-year-old hand ax is one of the first machines, a wedge-shaped cutter made by chipping away at a fist-sized stone.

The Pyramids of Giza were built around 2,500 BCE. At 481 feet (147 m) high, the Great Pyramid (far right) was the tallest structure in the world until 1311 CE when it was overtaken by Lincoln Cathedral!

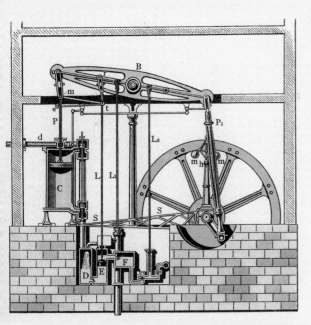

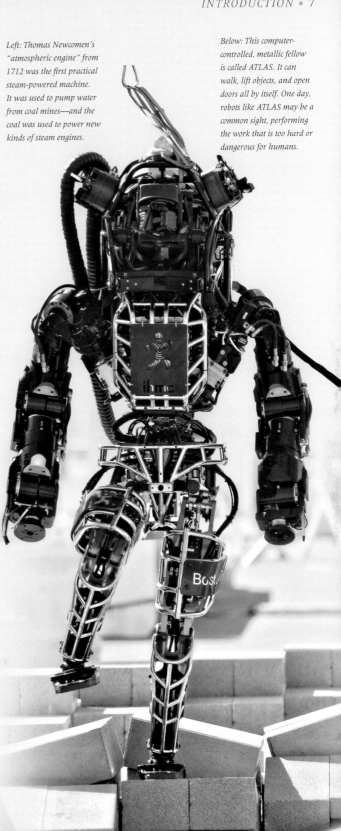

Left: Thomas Newcomen's "atmospheric engine" from 1712 was the first practical steam-powered machine. It was used to pump water from coal mines—and the coal was used to power new kinds of steam engines.

Below: This computer-controlled, metallic fellow is called ATLAS. It can walk, lift objects, and open doors all by itself. One day, robots like ATLAS may be a common sight, performing the work that is too hard or dangerous for humans.

t this point you might be saying "Industrial Revolution" to yourself, ut there was another that came before: The Agricultural Revolution, which began in the 18th century. New farming techniques developed in ngland along with new technologies, such as seed drills and improved low designs, made it possible for fewer people to grow enough food or the whole population. That created a workforce for new jobs—in the ctories of the Industrial Revolution.

O LOOKING BACK

he Industrial Revolution, which ran through most of the 19th century, as a huge period of change, when engineers developed ways of anufacturing goods on a large scale, and invented trains, cars, ocean-oing ships, and the first flying machines. What's happened since then? More of the same, much more.

Today, more than half of the world's population live in cities, not in the countryside. Cities, with their roads, sewers, power grids, and skyscrapers, are human habitats entirely created by engineers. In future, we may engineer homes in space, and develop robotic machines that think for themselves to do our jobs for us. All this technology requires power, raw materials, and creates pollution. How will we solve these very important problems? We should ask the engineers.

THE APPLICATION OF SCIENCE

ENGINEERS WORK IN MANY FIELDS, BUT IN EVERY CASE THEY ASK THE SAME QUESTION: HOW CAN WE APPLY WHAT WE KNOW TO SOLVE A PROBLEM OR IMPROVE OUR LIVES? LET'S TAKE A LOOK AT THE MAIN TYPES OF ENGINEERING. TOGETHER THEY MAKE THE MODERN WORLD—AND THE FUTURE.

AEROSPACE

Anything that moves through the air is designed by an aerospace engineer. That includes aircraft but also cars, trains, and turbines. Aerospace engineers research how air flows around their designs—the smoother the better.

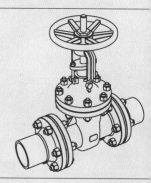

CIVIL

Roads, bridges, dams, and other features of the infrastructure are designed and built by civil engineers. They often work with concrete and steel and design structures that will stay standing for decades or more.

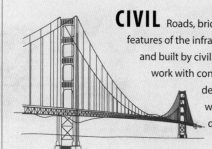

INDUSTRIAL

The machines and tools used in factories are built by industrial engineers. This is a complex field where several systems are designed to work together, making industry safer and more efficient.

ENVIRONMENTAL

This field looks for engineering solutions for problems of pollution and habitat damage. This includes developing renewable energy sources, such as solar power. In addition, geoengineers are looking for ways to reengineer the climate.

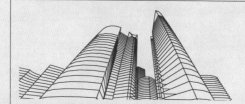

STRUCTURAL

One of the oldest arms of engineering, this field ensures that structures are strong enough to stay standing. Structural engineers work at all scales, from the design of a timber outbuilding to that of a cloud-piercing skyscraper.

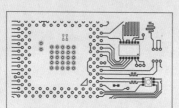

COMPUTER

Developing hardware (the electronics, displays, and input devices) and the software (the programs). All must be designed to work together.

MARINE

Sometimes known as naval architecture, this field of engineering handles anything that floats in or stands in water. As well as designing vessels that float safely and move efficiently, marine engineers work on the large engines used by ships.

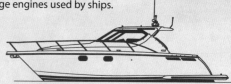

GENETIC

Where engineering meets biology, genetic engineers seek to create new forms of life by editing genetic material. This technique makes it possible to transfer the genes of one species into another—and may one day use entirely artificial genes.

ELECTRICAL

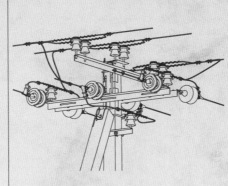

A huge arm of engineering that concerns everything from building power plants and a safe electricity supply network to designing useful electric machines, from toaster ovens to cars. A big challenge for electrical engineers is to develop new batteries or other systems for storing electricity.

CHEMICAL

This type of engineering is concerned with developing industrial processes that can produce useful chemicals from raw ingredients. Most chemicals are derived from petroleum but engineers are looking at alternatives, such as coal or even the gases in the air.

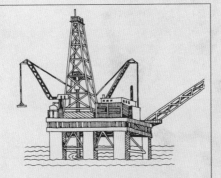

MECHANICAL

An ancient field of engineering that handles anything with moving parts. This includes simple machines, such as levers, wheels, and screws, as well as combining these machines to make more complex devices. It also includes developing engines, which turn heat energy into motion energy. Mechanical engineers find roles in many areas from car manufacture to the development of robots.

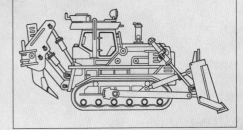

ROBOTIC

This kind of engineering integrates computing with mechanical engineering to build machines that are able to sense their surroundings and operate within them autonomously to carry out preset tasks.

MATERIALS SCIENCE

This field falls between engineering and science. Materials scientists are researching the properties of materials, testing out new alloys or plastics, to see if they can be used by engineers in improved designs. As well as testing strength, materials scientists research electrical and magnetic properties and investigate how materials respond to temperature changes and chemicals.

MEDICAL

Engineering has a major role in medicine. It is used to develop better diagnostic tools, such as the MRI scanner, and it also produces devices such as pacemakers, prosthetics, and drug pumps that help to improve the quality of patients' lives.

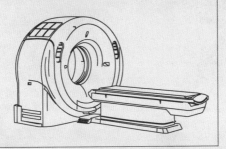

1 Stone Technology

ENGINEERING WITH STONE IS OLDER THAN THE HUMAN RACE. The distant ancestors of *Homo sapiens* (modern humans) began making and using stone tools as long as 3.3 million years ago.

Recent evidence suggests that Australopithecines, our earliest bipedal ancestors, used cracked rocks as cutters. However, the oldest stone tools that followed a design belong to the Oldowan toolkit, first discovered at Olduvai Gorge in Tanzania. It consists of hammerstones and cores. The cores were bashed on hammerstones to break off sharp flakes of stone that were used as cutting edges for slicing skin and meat.

Hand axes are usually 5 to 8 inches (12.5 to 20 centimeters) long and generally teardrop shaped. Larger axes were not used as tools, suggesting they were probably ceremonial objects or a sign of status.

Slow innovation

It took nearly another million years before *Homo erectus*, a closer ancestor to modern humans, learned to break off larger flakes and then sharpen them by cracking off, or knapping, smaller flakes from around the edges. This revolution in stone age technology was the beginning of the Acheulean toolkit, named after the site of St. Acheul in France where tools of this nature were first discovered in 1847. These implements included a tool made from core stone called a hand ax. The wear patterns on Acheulean hand axes show that they were used for various tasks including digging, cutting wood, butchering, and skinning animals. The new technology spread rapidly. The earliest known Acheulean tools from Africa have been dated to 1.6 million years ago, and those found in South Asia and Europe are from just a few thousand years later.

New techniques and materials

The Acheulean tool design was incredibly long lasting, continuing to be made in some places until around 250,000 years ago, around the time the modern *Homo sapiens* species emerged in Africa. Then further technological developments were made, with more emphasis placed on tools made from small flakes rather than the larger cores, while bone and antler tools became more common.

KNAPPING AND LANGUAGE

The evolution of language is a fiercely debated topic. Recent studies have found evidence that the ability to make tools and the ability to communicate may have evolved together. Brain scans of modern craftsmen knapping stones using ancient techniques show that they are using the same part of the brain that is linked with speech. Perhaps language evolved as our ancestors taught the techniques of stone technology to their children.

This barbed harpoon point carved of bone was made in Africa 85,000 years ago.

2 Taming Fire

IT IS IMPOSSIBLE TO SAY EXACTLY WHEN OUR ANCESTORS first began to control and use fire. The ability to control fire was a major step forward in the history of human civilization. It gave light, warmth, protection, and a better diet.

Some open-air archaeological sites might indicate that our ancestors were controlling fire 1.5 million years ago. But these sites could be from natural fires. Recent findings suggest that the first hominins (humans or our ancestral species) to make deliberate use of fire were *Homo erectus* living in Africa over a million years ago. Researchers investigating South Africa's Wonderwerk Cave have uncovered the remains of campfires with charred animal bones, suggesting that the fires were used for cooking. (The earliest confirmed constructed fireplace was found in Qesem Cave in Israel dating from 300,000 years ago.)

The Wonderwerk Caves added fuel to the "cooking hypothesis," a theory of the origins of humanity as proposed by zoologist Richard Wrangham. When *Homo erectus* appeared around 1.8 million years ago it was radically different from the likes of *Homo habilis* that came before, with much larger brains, smaller teeth, and bodies that were very similar to ours. Wrangham suggested that the drive behind the transformation was fire. Cooking food made it easier to eat and digest, providing the extra calories for larger, energy-hungry brains, and the cooking fires had the added bonuses of providing warmth and warding off predators.

A Native American bow drill is an ancient firelighting tool. The long spindle was twisted back and forth repeatedly in a slot on the hearth board (hanging in this image). The friction between the two hard, wooden objects created enough heat to ignite kindling.

MATCH

It was a long time before anyone invented a reliable, portable, self-igniting source of fire. In 1826, John Walker, a chemist from Stockton-on-Tees in England, tried to scrape off a mixture of chemicals (including sulfur and phosphorus) that had dried on to the end of his stirring stick. To his astonishment, when he drew the stick across the stone floor of his laboratory it burst into flame. Completely by accident, Walker had invented the friction match. He was soon selling "Friction Lights," packaged in a cardboard box complete with sandpaper for striking.

Sparks and friction

Our ancestral hominins first used fire ignited by lightning. Only later did they develop firelighting technology. There are two techniques: Percussion, where sparks are made by hitting stones together, and friction, where heat is produced by rubbing hard surfaces until flammable material gets hot enough to burn.

3 The First Boats

THE OLDEST BOAT EVER FOUND IS THE 10-FOOT (3-METER) LONG PESSE CANOE, found in Holland and was constructed around 8,000 BCE. But as old as it is there is much evidence that people had taken to the water as long as 800,000 years ago.

Homo erectus in Africa could make stone tools and tame fire. They also learned how to make boats. During the 1.5 million years that they existed, *Homo erectus* spread out of Africa and around the world. Their stone tools have been found in the Indonesian archipelago on the island of Flores, which cannot be reached from the mainland other than by many island-hopping sea crossings of up to 10 miles (16 kilometers). This is compelling evidence that *Homo erectus* was building seaworthy craft 600,000 years before the appearance of modern humans.

No trace remains of these early boats but, based on the tools and materials available to the boatbuilders, the best guess is that bamboo rafts were used. Probably differing little from these prehistoric ones, bamboo rafts were widely used throughout Asia until the late 20th century.

The dugout canoe was a step up from primitive rafts. Although rafts could be made using simple cutting tools, a dugout required more substantial tools, like axes and chisels.

Ocean going

Around 6,000 years ago, East Asian migrants spread through Southeast Asia and into the Indian and Pacific Oceans by making the first long-distance ocean voyages. These were made in canoes with an outrigger, a float extending to the side, which made the boat stable in rough seas. Outrigger boats were used to steadily populate ocean islands, only reaching New Zealand, the last significant uninhabited landmass, around 700 years ago.

SAILING

The first known sailing vessels appeared in Mesopotamia around 5,000 years ago. Their square sails offered little control and carried the vessel in whichever direction the wind was blowing. The triangular "lateen" sail, as seen on this Indian Ocean dhow, was invented in the Mediterranean about 2,200 years ago. It can sail against the wind and is used in modern sailboats to this day.

Anything that floats can form a serviceable craft. This reed boat is a traditional design of river-boat used in Mesopotamia A similar design is also seen in Lake Titicaca in th Andes Mountains.

4 Pottery

AROUND 20,000 YEARS AGO PEOPLE BEGAN TO DISCOVER how to change the properties of the materials they used. Made from clay, pottery, or ceramics, is the first instance of humans creating a synthetic material.

Clay is a very common and abundant natural material and is easy to shape. Probably the art of pottery-making was independently discovered in many places. The earliest examples of clay usage that have been found are figurines dating from 29,000 to 25,000 BCE that were unearthed in the Czech Republic. Archaeologists reckon that people began using clay to make functional vessels for holding water or food around 13,000 years ago in Japan. It is perhaps no coincidence that this occurred at the same time that settled agricultural communities were being established: Hunter-gatherer tribes have little use for lugging around heavy—and fragile—clay containers.

Developing techniques

Probably people noticed that wet clay went hard in the heat of the sun and wondered if this natural process could be improved upon. As far back as 13,000 years ago Japanese potters "fired" their clay by simply placing the object in an open fire (see box below).

These first pots were made by building and smoothing coils of clay until the desired size and shape were achieved. Making a coiled pot is laborious. The potter has to walk around the growing pot, squeezing and smoothing the clay as it is layered to ensure evenness. It would be easier if the pot could be made to turn. The invention of the potter's wheel dates back to around 3500 BCE in Mesopotamia, about the same time as the wheel was first used for transportation. The first potter's wheels were simply small turntables for faster coiling. It would be several more centuries before the fast wheel and the "throwing" technique—shaping a lump of clay on the wheel—appeared.

Many of the earliest pottery items are small figurines, such as this one from Bulgaria that was made about 7,000 years ago.

...ottery from the Jomon ...lture of Japan is thought ...be the oldest example of ...ed clay vessels. This one ...as made by shaping and ...oothing the clay ...hand.

Fine china

By 3,000 years ago Chinese kilns could reach 2,730° Fahrenheit (1,499° Celsius). At this temperature, clay becomes glassy, so pottery objects could be made thinner and lightweight. This material became known as "china," or porcelain.

FIRING

When clay is heated to at least 1,832° Fahrenheit (1,000° Celsius), the minerals dry out and melt, fusing into a more rigid form. Known as firing, this is most simply done by burying pots inside a fire (right). But a kiln, an oven specially made for the purpose, can reach higher temperatures for longer periods. Before firing, pots are generally glazed, or coated in salts or ash. This makes the fired pottery waterproof and means the finished item can be painted.

5 Megalith

MEGALITHS (MEANING "BIG STONES") WERE USED to build monuments between 4500 and 1000 BCE. The effort involved in the construction of these structures shows they were of considerable significance to the people who built them.

One of the most common types of megalithic structure is the dolmen, a small chamber built under huge, cut stones. There are generally graves (probably of important people) inside and around dolmens. Found throughout Europe, Asia, and Africa, the typical dolmen is formed from upright supports with a flat capstone on top. These capstones can exceed 100 tons. The oldest known dolmens, found in Western Europe, are around 7,000 years old. There is much discussion as to how these heavy capstones were raised into position. One idea is that they were laboriously hauled up earth ramps. Another possibility is that they were levered upward, lifting one side a few inches and chocking with timbers, and then doing the same with the other. Finally, the timber chocks were replaced by upright support stones.

Stonehenge in southern England is the world's most famous megalithic structure. Its stones were erected about 2,500 years ago, but researchers are still finding older and younger structures made of stone, earth, and wood, in the area.

Standing stones

Another megalithic structure is the menhir (meaning "long stone"). This is an upright stone, often of great size, which was frequently placed with others in lines or circles. The Carnac site in France has 2,935 menhirs arranged in parallel rows dating from 6,500 years ago. Stonehenge in England famously features another type of megalithic structure called the trilithon. This consists of two upright stone posts topped by a third stone—the lintel—set horizontally across them. Menhir structures obviously had a ceremonial function, but exactly what is an intriguing area of debate.

WORK OF GIANTS

Many legends have sprung up to explain how megalithic structures came to be built. Stonehenge, for example, was said to have been constructed by the wizard Merlin, who recruited a giant to carry out the work. Dolmens in Portugal were the work of Enchanted Moura, a beautiful, red-headed bull-woman with hooves. Moura could also weave the rays of the sun and in addition she is credited with teaching people how to spin, weave, plow, and brew beer.

6 Building Cities

OUR WORD "CITY" COMES FROM THE LATIN *CIVITAS*, from which we also get civilization. Cities mean civilization and people began to build permanent settlements when they started farming around 12,000 years ago.

The advent of agriculture meant that food was more readily available and could be stockpiled for months in advance. Freed from the necessity to spend most of the day sourcing meals, people could divert their time to other things and specialize in particular skills—with some becoming the first engineers!

Ancient remains

The settlement of Çatalhöyük in Turkey shows what life in the first cities was like. The houses there date from 9,500 years ago and had large, plastered rooms with raised platforms for use as tables and beds. It appears that residents worked on the roofs during good weather, and the roofs of neighboring houses merged into a communal plaza.

Uruk, in the Tigris–Euphrates valley in what is now Iraq, is the oldest city for which there is documentary evidence. It was first settled around 4500 BCE. However, Aleppo, a modern city in Syria, may have been settled in 6000 BCE, and Jericho in the West Bank of the Palestinian Authority could be as old as Çatalhöyük.

At Çatalhöyük in Turkey, the remains of a city built around 9,500 years ago have been uncovered, showing us what homes were like at the start of the Bronze Age.

e buildings in alhöyük were all same size, and so haeologists have posed that the buildings re communal, with some d for worship (above) d others for domestic poses.

City plan

By around 5,000 years ago, cities habitually had defensive walls. The center of the city would likely be dominated by the ruler's palace and a temple to whichever deity the city favored. Houses were crowded together around this central area. Larger populations required social organization and some sort of infrastructure to keep things running smoothly, and that in turn required building structures for carrying out various essential tasks, such as storehouses for surplus food or workshops where tools could be made.

BRICKS

Evidence suggests that Jericho's buildings were made of mudbricks, which were used as early as 8350 BCE. These are simple to make, giving good use of the waste material left over from the grain harvest—just mix mud with water, add straw as a binder, pack into a mold, and leave to dry in the sun. Brick has been a dominant building material ever since, mostly made today from baked clay.

7 The Plow

THE PLOW IS ONE OF THE MOST IMPORTANT AGRICULTURAL IMPLEMENTS ever developed, helping to break up and aerate soil to plant crops, incorporate fertilizer, and manage weeds.

The ancestor of the plow was the simple digging stick, a sharpened branch used to break the ground prior to planting. Eventually, handles were added to the digging

stick to make it easier to push through the soil. Around 6000 BCE, the domestication of the ox in Mesopotamia and the Indus Valley provided a new source of power for working the fields. The scratch plow, or ard, was the first design. It was a digging stick fixed to a yoke that could be harnessed to a

A wooden plow from the Middle Ages.

draft animal. The ard was only good for making furrows in light, sandy soil. Around 1000 BCE the iron plowshare began to be used, and gradually replaced the ard. The plowshare uses a sharp cutting edge to slice through wet and thick soils.

8 Metalworking

IT IS IMPOSSIBLE TO OVERSTATE THE CRUCIAL ROLE that metalworking has played over the last 8,000 years of human history. Metal materials can be worked into any shape and can be formed into vitally useful hard and sharp objects.

The first metals used were silver, copper, and gold, all of which can be found in their metallic, or native, states in nature. Although copper does sometimes occur in almost pure nuggets, like most other metals it is usually found in combination with other materials as part of a rocky ore. It takes high temperatures to separate copper or other metals from their ores (a process called smelting), and it is fairly likely that the first people who discovered how to do this were potters experimenting and trying out new techniques for firing their ceramics. It is tempting to imagine an unknown potter watching with interest as a bright rivulet of molten metal snaked out from his kiln.

GOLD

Gold does not readily combine with other elements, so does not mix with oxygen to oxidize (tarnish or corrode), a fact that makes it ideal for creating jewelry and other art. This incorruptibility, combined with its rarity, made gold a symbol of power and prestige. Around 2,500 years ago, the goldsmiths of King Croesus of Lydia in modern Turkey developed improved refining techniques that permitted the king to establish the world's first standardized gold currency. Gold has been associated with wealth ever since, as an intrinsically valuable material.

Copper to bronze

Between around 5000 to 3000 BCE, the copper trade grew
in importance among the peoples of the Near East and the
Mediterranean. Copper is a soft metal, good for making jewelry but
not much use for making tools or weapons.

Copper is often found in an ore mixed
with tin, and when the two were
smelted together they formed
a tougher alloy, or mixture
of metals. Mixing copper
and tin in a ratio of about nine
to one makes the alloy bronze. Bronze
was the first industrial metal. It is
much harder than copper or tin
separately, and it can be beaten
into a cutting edge. It is also easier
to melt than copper, so therefore it
is easier to cast in molds.

The Bronze Age, the time in history when
bronze was the main material in general use,
lasted for roughly 2,500 years from around 3500
to 1000 BCE. It was a time when civilizations
rose and fell and important trade routes were
established, such as those carrying the tin that
was essential for making bronze.

Hotter and stronger

Around 1200 BCE, several discoveries combined to
usher in a new age of metal—the Iron Age. First,
charcoal was used for fuel, which released some
impurities from iron ore, then bellows were invented,
which increase the amount of oxygen in a furnace and
therefore obtain the higher temperatures needed to
extract iron.

Known as slag pits or bloomer furnaces, these
ancient kilns still weren't hot enough to fully melt
iron. Instead, they produced a bloom, a mix of iron
and other materials that could be refined by repeated
heating and hammering. Iron is the fourth most
common element on Earth, and succeeded bronze as
the metal of choice not because it was of better quality,
but because it was easier to make in large quantities.
As furnaces improved so did the quality of iron, which
is stronger and harder than bronze.

*...key to a successful
...lter is high temperature
...pled with the presence
...arbon. The burning
...bon reacts with the
...per, tin, or iron ore,
...oving the impurities
...n the metal.*

*This Bronze Age helmet
from Sparta was designed
to protect the wearer's head
from swords and spears
also made from bronze.*

9 The Wheel

TRY TO IMAGINE WHAT THE WORLD WOULD BE LIKE if there weren't any wheels in it. It's next to impossible. The wheel is without doubt a contender for the greatest world-changing invention of all time, used in transportation, pottery, clockwork—and even as a computer disk!

A chariot appears as a detail in a Sumerian painting of soldiers from 2500 BCE.

A great many of our engineering innovations were inspired by nature, but not the wheel. Natural selection has yet to produce a single wheeled part in the living world. The available evidence seems to point toward the wheel first being used in Mesopotamia over 5,000 years ago as an aid to potters before someone had the bright idea of using it for transportation.

The breakthrough was not just the wheel, but the wheel and axle. Early wheels were fixed to the axle so both turned while attached to a wagon. Nevertheless, this required a degree of precision. The wheel must be well rounded, and the axle has to be the right size: Too thick and the wheel won't turn freely; too thin and it won't be able to bear a load. The first wagons were narrow with short axles so they could be as thin as possible.

Rollers were the predecessors of wheels as methods of transporting heavy loads. They were not linked to the load and had to be moved from the back to the front to continue the journey.

Birthplace of the wheel

The invention of the wheel and axle was such a huge challenge that in all likelihood it happened only once but then spread. Whoever invented it must have had access to metal tools, which would have been needed to shape the wheels and axles. The earliest images of wheeled carts have been discovered in Poland and elsewhere in central Europe and there is speculation that while Mesopotamia may have had the potter's wheel the most likely birthplace of the wheel for transportion was eastern Europe. One possible contender is the Tripolye people of Ukraine: Wheeled toy wagons made by them have been found dating back to around 3800 BCE, and may well have had full-size counterparts.

DRAGGING LOADS

The travois, or drag sled, was used by Native Americans and many other cultures in lieu of the wheel. It consisted of two wooden poles, with a platform suspended between them, which held the load. This was attached to the back of a dog or horse. Often this was done by simply hanging a pair of tepee poles across the horse's back and attaching a baggage platform between the poles trailing behind.

10 Rope Stretchers

CONSTRUCTION ENGINEERING REQUIRES accurate surveying of land. In ancient Egypt surveyors were called *harpedonaptai*, or "rope stretchers."

The rope stretchers' job was to measure the boundaries of fields. They did this by stretching out a rope with knots at equal distances along it and counting the knots. They also used the plummet, a simple lead weight fixed to an A-frame, for determining vertical lines. The corners of the fields were plotted out using a triangle of ropes with three, four, and five knots on the three sides. These numbers are now known as Pythagorean triples, which means a triangle with these lengths will always form a right-angled triangle. Having the ability to make accurate right angles ensured the plots for long fields were also accurate and was immensely useful when carrying out monumental building projects, such as the pyramids.

...des with three, four, and ...e knots (counting one ...rner tie) always produce ... right-angled triangle.

An Egyptian rope stretcher at work in the fields of the Nile Valley. The rope stretcher was appointed by the pharaoh to ensure there were no disputes over farmland.

11 Ziggurat

FROM AROUND 2500 TO 500 BCE THE PEOPLE OF MESOPOTAMIA took to building huge, stepped temples made of brick in their cities.

The ziggurat itself had no internal structure; the gods of Mesopotamia are often linked with mountains and the suggestion is that ziggurats were built to mimic the homes of the gods in the heart of the city. Around 25 ziggurats are known today, none of which survive to their original height. For almost half of these no means of ascending to the temple has been discovered; for the others the ascent was made by spiral ramps. The biblical Tower of Babel is thought to have been based on the ziggurat of the temple of Marduk in Babylon. Similarly, the Hanging Gardens of Babylon are likely a reference to the practice of landscaping the sides and terraces of ziggurats with trees and shrubs.

The ziggurat in Chogha Zanbil, Iran, is the largest example still standing. It is 344 feet (105 meters) wide and 80 feet (24 meters) high today, perhaps half its original height.

12 Pyramid

THE PYRAMIDS OF ANCIENT EGYPT WERE BUILT OVER A PERIOD spanning 1,000 years to house the bodies of the pharaohs. Pyramid-building stopped around 1800 BCE when it was finally realized that they were simply a magnet for graverobbers and offered no real protection for the ruler's remains. After this the kings and queens were buried in secret tombs.

INSIDE THE GREAT PYRAMID

Perhaps surprisingly, we have still not uncovered all there is to find inside the pyramids. Working under strict guidelines to avoid damaging these historical artifacts, researchers have only gained access to a few chambers inside the Great Pyramid at Giza. These include the King's Chamber (near the center of the pyramid), the Queen's Chamber (slightly lower down), and the Great Gallery, which connects the two chambers. There is also a chamber, seemingly abandoned during construction, that is below ground level. New technology, such as infrared thermography and robot crawlers, is being used to look for more details, but the idea of breaking into tunnels and chambers is seen as irresponsible vandalism today.

According to Egyptian tradition, the remains of the dead had to be left unsullied if they were to make the journey to the afterlife. As a result great efforts were made to build secure mausoleums for Egypt's rulers. The early Egyptian kings were buried in a type of tomb called mastaba. These were flat-topped rectangular structures of mudbrick or stone with a shaft descending to a burial chamber deep underneath. Around 2780 BCE, Imhotep, architect to King Djoser, had the idea of placing six mastabas, one on top of the other, to form a stepped pyramid. Within it were a number of rooms and passages, including the king's burial chamber. Imhotep's most ancient pyramid still stands on the west bank of the Nile River at Sakkara near the ancient city of Memphis.

Flat sides

The next step in the development of the pyramid, from stepped to smooth sides, happened during the reign of King Snefru (2613–2589 BCE). A step pyramid was built at Medun, which was then filled in with stone, and covered with a limestone casing. Work was also begun on a pyramid that was apparently planned to have smooth sides. About halfway up, however, the angle of incline decreases by several degrees, and the sides rise less steeply, the unusual shape earning it the name of the Bent, or False, Pyramid. It's thought that the architects decided on the change in angle to make the pyramid more stable. The earliest tomb known to have been designed and built as a true pyramid is the Red Pyramid at Dahshur, which may also have been erected by Snefru. It is about 722 feet (220 meters) wide at the base and 344 feet (105 meters) high.

MESOAMERICAN PYRAMIDS

Beyond the Egyptian pyramids, those of Central and South America are perhaps the best known. They include the Temple of Kukulcan in Chichen Itza, Mexico (below), and the Pyramids of the Sun and Moon, which dominated the ancient city of Teotihuacan, near today's Mexico City. When the Pyramid of the Sun was completed around 200 CE, it stood 216 feet (66 meters) tall and 760 feet (232 meters) wide—one of the largest structures built in the pre-Columbian Americas. Beneath the pyramid a cave leads to a set of chambers, apparently the site of various rituals.

The largest and most famous of all the pyramids is the Great Pyramid at Giza. This was built by Snefru's son, Khufu. The pyramid measures 756 feet (230 meters) along each side of its base and originally stood 481 feet (147 meters) tall, although today, stripped of its outer casing of white limestone, it is now just 455 feet (139 meters) high. Researchers have estimated that the stone blocks used in its construction average over two tons apiece, with the largest weighing as much as fifteen tons.

Construction system

Just how exactly the pyramids were constructed has been the subject of considerable speculation. Water-filled trenches were used to make sure that the construction site was properly leveled. Rope stretchers ensured that the square base of the pyramid formed accurate right angles, while the Egyptians' knowledge of astronomy allowed them to orient the pyramids to the cardinal points of the compass.

The immense stone blocks used in building the pyramids were dragged onto the site on wooden sledges, the ground having first been wetted to reduce friction. They were then hauled up ramps into position on the pyramid.

The Egyptian builders cracked stone from quarries using dry wooden wedges that then expanded when soaked with water. The stone was then shaped with copper tools. However, copper would have been too soft to cut the hard granite used in parts of the pyramid. It is possible that the stonemasons used sand as an abrasive to coat their drills and saws. In any case, it would have been lengthy and laborious work. According to the Greek historian Herodotus, writing in the fifth century BCE, 100,000 men were employed for three months a year for twenty years to build the Great Pyramid, although modern researchers believe the number to have been much smaller.

It is generally assumed that the blocks used to make a pyramid were hauled into place along earthwork ramps. However, the exact structure of these ramps is still not clear. Perhaps they were spirals that surrounded the stonework, or perhaps they were vast inclines.

Giza is now home to three large pyramids. In this view the Great Pyramid (the largest) is at the back, then comes Khafre's Pyramid (the son of Khufu), and the smaller one in the foreground was built for Menkaure.

13 The Arch

ONE OF THE GREATEST ACHIEVEMENTS OF ROMAN TECHNOLOGY was the use of the arch. Arch building began as early as 1800 BCE but ancient Egyptians and Greeks considered them unsuitable for their architecture.

It was the Romans who perfected the arch, using them in bridges, aqueducts, and large-scale buildings such as the Colosseum. Arches have the advantage over lintels (a horizontal bar across an opening) in that they direct the weight of the building above them sideways. While a flat lintel will crack if supporting a wide span, a curved arch directs the weight into its supports and so is considerably stronger. The essential component in arch construction is the wedge. The wedge-shaped blocks from which an arch is formed are called voussoirs. Each one is cut precisely to fit firmly against the surface of its neighbors. The central voussoir is called the keystone. While the arch is being constructed the voussoirs have to be supported; this is generally done using timbers, which are removed once the keystone has been set in place.

When completed, the Colosseum in Rome had 240 arches in three stories.

14 Irrigation

THE FIRST CIVILIZATIONS IN EGYPT AND MESOPOTAMIA were built near rivers which supplied water for crops. Settlements based far from easily accessible waters had to develop new methods of irrigation.

The ancient Egyptians employed basin irrigation, which made use of the regularity of the Nile floods each summer. This involved holding back the floodwater in ponds, some as big as 50,000 acres (20,235 hectares), for as long as possible to allow the sediment to settle. The water gradually drained away over a period of weeks as the river level fell, leaving behind a deposit of fertile silt. Farmers sowed their fall and winter crops in the waterlogged soil. This system only allowed a single crop each year, and the farmers were at the mercy of the floods.

The floods in Mesopotamia, fed by the Tigris and Euphrates rivers, were less predictable, and liable to be severe when they

SHADUF

The shaduf was developed in Egypt and Mesopotamia and is still used in many countries. It consists of a pole mounted on a fulcrum with a counterweight on the short end and a bucket suspended on a rope from the long end. The operator pulls down on the rope to lower the bucket into the river and allows the counterweight to raise the bucket back up again, lifting water with it.

did come. The Sumerian engineers developed a means of trapping the floodwaters and distributing the water to the fields through a series of small channels. The drawback of the Sumerian system was that it led to an accumulation of salt in the soil, which consequently lost fertility as a result.

Around 1000 BCE the Persians (in present day Iran) developed the qanat, a water-supply system that is still used in some of the world's dry regions. The qanat taps underground mountain water sources and channels the water downhill through a series of gently sloping tunnels, which can often be several miles or kilometers long. An advantage of this is that the underground tunnels prevent evaporation of the water. Additionally, as the flow is powered by gravity, there is no need for pumping. Qanat tunnels were excavated by hand with vertical shafts sunk at regular intervals to remove material from the tunnel and to provide ventilation.

The main qanat tunnel was just large enough for a person to travel through. It opened out into canals which distributed water to fields for irrigation.

15 Trireme

THE FAST, MANEUVERABLE TRIREME WAS THE PREMIER WARSHIP of its era, named for its three decks of oarsmen. The naval power of its trireme fleet enabled the Athenians to dominate the Aegean Sea in the 5th century BCE.

The trireme's name comes from the arrangement of rowers in three tiers of as many as 30 oars about 15 feet (4.5 meters) long ranged along each side of the ship. A full complement of rowers could achieve speeds as high as nine or ten knots in short bursts. The trireme was also equipped with two sails of papyrus or flax, used when cruising but taken down and stored when the ship went into battle. The remains of boathouses that have been discovered suggest that the maximum length of a trireme would have been around 121 feet (37 meters). The ships were built using oak for the outer hull and pine, fir, and cypress for the interior. The trireme's principal weapon was its bronze-sheathed battering ram, which was driven into enemy vessels; occasionally, this might result in a sinking but usually it was done prior to boarding the enemy ship.

Triremes were powered by sail for long journeys, but during battles, the oarsmen took over to drive their vessels at ramming speed toward the enemy.

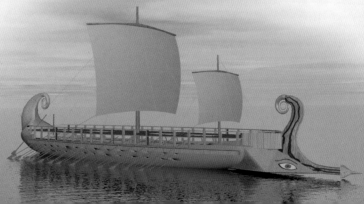

16 The Cloaca Maxima

ROME'S CLOACA MAXIMA WAS THE ANCIENT WORLD'S LARGEST SEWER. The outfall of the sewer into the Tiber river can still be seen today. Constructed in the 6th century BCE, it began as an open drain built to transport waste to the river.

Over the next few centuries the Cloaca Maxima—meaning "greatest drain"—was enlarged, extended, and roofed over to handle the increasing demands made on it. Wastewater from bathhouses, latrines, fountains, and public buildings, as well as rainwater runoff from the roads, all flowed into the Cloaca Maxima. Large drainage openings along the sides of the streets carried excess aqueduct water into the drainage system. The public would dump their wastes into these openings to be flushed away. The tunnel of the Cloaca Maxima ran for around 3,000 feet (900 meters) and was 15 feet (4.5 meters) wide and 12 feet (3.7 meters) in height. At the time of the emperors, around the first century CE, it was possible to travel through it by boat. The tunnel was covered by a vaulted roof formed from massive squares of limestone, with arches inserted at roughly 13-feet (4-meter) intervals.

The sturdy arches of the Cloaca Maxima's outflow have survived while many of Rome's ancient buildings have not. The sewer often allowed floodwater from the river into the city, but in the 20th century, the Roman sewer was connected to the modern drainage system to prevent this from happening.

17 Icehouse

BEFORE THE INVENTION OF REFRIGERATION, it wasn't easy to get a cool drink or to store perishable foods, especially in the hotter parts of the world. One answer was to build an icehouse.

The invention of the icehouse stretches surprisingly far back into history. There is evidence that one was built in Mesopotamia as early as 1780 BCE. The remains of icehouses dating back to 700 BCE have been uncovered in China and the likelihood is that they were in use much earlier than that.

Making ice
The most common way of trying to preserve ice for as long as possible was by insulating it. The icehouse was often an underground chamber, its precious contents packed in straw

SHARBAT
The terms sherbet and sorbet both refer to fruity ice treats. Both the words are derived from the Persian *sharbat* (*sharab* in Arabic), which is a sweet, chilled drink, traditionally flavored with fruits or flower petals. The drinks were developed in Persia at least 2,000 years ago thanks to the prevalence of fresh ice and introduced to Europe around 1,000 years ago by the Moors. In Sicily, the *sharbat* became granita, a fruit-laced ice desert. In Italy and later France, the iced dish evolved into sorbet before the rise of ice cream proper—a mix of milk fat and sugar—which was developed in the 1600s.

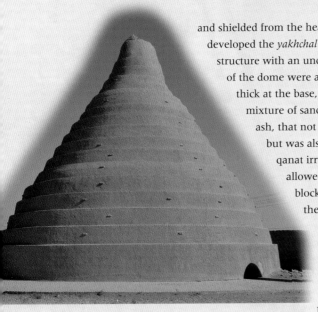

and shielded from the heat of the day by thick walls. The Persians developed the *yakhchal* (Persian for ice pit). This was a domed structure with an underground storage space inside. The walls of the dome were at least six and a half feet (two meters) thick at the base, and made of mudbrick coated with a mixture of sand, clay, egg whites, lime, goat hair, and ash, that not only had excellent insulating properties but was also waterproof. In winter, water from the qanat irrigation system was fed into channels and allowed to freeze overnight. The ice was cut into blocks and stored in a yakhchal, which was then sealed.

The cone shape of a yakhchal helped keep the ice cool. Warm air rose inside the dome and whistled out through vents in the top. The external wall of the yakhchal was covered by straw during the day to protect it from the heat of the sun and at night the straw was removed to allow heat from the yakhchal to radiate out into the cold night air.

Catch the wind

The Persians also built wind towers called "badgirs," or windcatchers. This is a sort of passive air conditioning system that works by redirecting the flow of the wind, trapping it in vents above the roofs of buildings and channeling it down to cool the rooms below. One common type of wind tower draws air out of the building so it can be replaced by air cooled by an underground qanat. The wind tower is opened on the opposite side from the prevailing wind and the pressure differential on one side of the building causes air to be drawn down into a passage on the other side down to the qanat. The hot air coming in from outside is cooled by the cold water running through the qanat. Then the cooled air is drawn back up through the windcatcher, cooling the building as it flows through it. A more basic design of windcatcher was also used in Egypt to keep houses cool. Clay models of houses with windcatchers have been found dating back to 1100 BCE.

Windcatchers on a traditional Persian building. The vents face different directions and can be opened and closed to catch the prevailing winds.

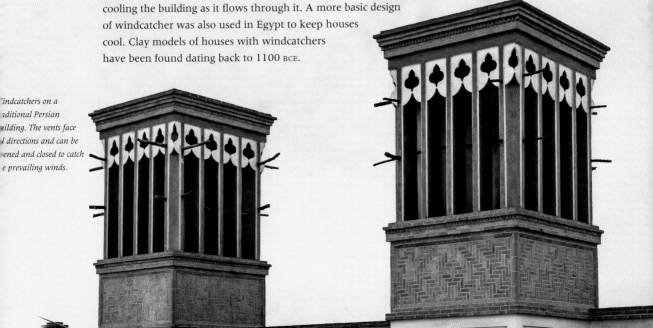

18 The Parthenon

THE TEMPLE OF THE PARTHENON, STANDING ON THE ACROPOLIS, or citadel, of Athens, was built between 447 and 432 BCE and still dominates the present day city. The Acropolis hill covers around 7.5 acres (3 hectares) and rises 500 feet (150 meters) above the city.

The Parthenon was raised as part of a project to rebuild the Acropolis following an attack by the Persians in 480 BCE. From parts of the financial accounts that have survived, the construction budget for the Parthenon has been variously estimated at from 340 to 800 silver talents—quite a sum of money, considering that a single talent could pay the wages of a trireme crew for a month. Sophisticated architectural techniques were used in the building of the Parthenon, including making the columns lean slightly inward to make them seem perfectly straight from a distance. The architect knew that straight lines on this scale appear to lean inward from a distance.

The Parthenon is one of the most famous ancient buildings in the world. It was designed by the Greek artist Phidias.

19 Waterwheel

WATERWHEELS SPUN BY A RIVER CURRENT were first used for lifting water in Egypt in the 4th century BCE. Two centuries later, Greek and Rome engineers developed the waterwheel as a power source for other machines.

The rotation of a waterwheel is transmitted to the axle or spindle, which also spins. This motion can be used to drive simple machinery, most often in early days a mill for grinding flour, and later pumps and spinning machines.

There are two basic designs of waterwheel: Horizontal and vertical. The horizontal waterwheel consists of a vertical spindle with a set of paddles

fixed around the bottom and a millstone at the top. It is not very efficient, converting only about 15 to 30 percent of the kinetic energy of the water flow into usable energy. It needs a small volume of water, but the water has to be moving at a high velocity to turn the wheel. Horizontal waterwheels are suitable for use in areas where there are steep slopes with fast running water, like mountain streams.

Under and over

There are two principal types of vertical waterwheel: The undershot and the overshot. In the undershot wheel the lower part of the wheel is submerged in the current, which pushes directly against the paddles. This is slightly more efficient than a horizontal wheel because only part of the wheel is submerged at a time, which reduces drag forces that slow its rotation.

The overshot wheel has a channel, called a pentrough, that directs the water so that it falls on to a series of blades near the top of the wheel, thus causing the wheel to turn. The overshot wheel is far more efficient than the undershot wheel, achieving an efficiency of 60 percent.

Lifting water

The noria is a type of vertical wheel that is used for irrigation rather than mechanical work. Paddles fixed to the wheel utilize the energy of the water to turn the wheel while separate containers scoop up water with each turn. At the top of the revolution the containers spill their contents into a canal or aqueduct to be directed where it is needed. Wheels with wooden containers were used around the 2nd century BCE but the Romans introduced ceramic containers around the 3rd century CE. Norias are still used in Syria today.

TREADMILL

Technologically innovative though they were, the Romans were perhaps not quite as inventive as they might have been. One reason for this was the availability of slaves as a power source. The Romans developed a crane that could lift weights of 3 tons or more, powered by slaves pushing round a treadmill akin to a giant hamster wheel. Slave-powered waterwheels were also used to drain water from mineshafts. During the slave era in the Americas, treadmills run by slaves were a common power source, and in Europe they were powered by prisoners working in shifts.

An unused wooden overshot waterwheel has a wooden pentrough above that carried the water supply over the wheel.

20 The Pont du Gard

THE PONT DU GARD IS ONE OF THE MOST ACCOMPLISHED works of Roman engineering. Constructed in 19 BCE, this three-tiered aqueduct was built to carry water to the city of Nimes across the Gard river in southern France.

At 161 feet (49 meters), the Pont du Gard is the tallest Roman aqueduct still standing. It is all the more impressive given it was built without cement. It has not been used to carry water for 1,600 years, but is used as a river crossing to this day. The first Roman aqueduct was the 10-mile (16-kilometer) Aqua Appia built in 312 BCE to carry water into Rome. Many more were built, especially in the first century CE, as the empire grew. One particular innovation, the inverted siphon, allowed aqueducts to cross valleys. These were clay or lead pipes that used the force of gravity to build up pressure as the water ran down the valley, gaining sufficient force to drive it up the opposite side.

The aqueduct rises to a height of 160 feet (49 meters) above the river; s arches, ranging from 50 to 79 feet (15 to 24 meter wide form the first tier, w the widest arch spanning the river, eleven equal-siz arches make up the secon tier, and the third, carryi the water channel, is composed of 35 smaller 1 feet (4.5-meter) arches.

21 Simple Machines

MACHINES ARE TOOLS, SOMETHING THAT MAKES IT EASIER TO PERFORM A TASK. Every kind of machine, no matter how complex, can be seen as a collection of simpler devices that have been used by humans since the dawn of civilization.

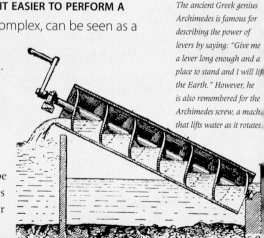

A simple machine directs a force from one place to another: A force applied to one part of the machine, called the effort, moves another part of the machine that overcomes a resisting force, called the load. The direction of the force may be altered and its size can be magnified (or reduced). There are six simple machines as described by Heron of Alexandria, a Greek engineer who lived in the 1st century CE. They are the lever, screw, ramp, wedge, wheel and axle, and pulley.

The ancient Greek genius Archimedes is famous for describing the power of levers by saying: "Give me a lever long enough and a place to stand and I will lif the Earth." However, he is also remembered for the Archimedes screw, a mach that lifts water as it rotates.

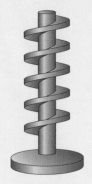

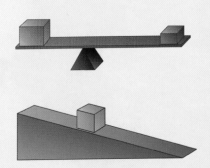

...ulley is a rope looped ...und one or more wheels. ...its simplest, a pulley ...irects the force applied to ...rope, but adding more ...eels will magnify the ...ce making it easier to lift ...avier loads.

The wedge focuses a force applied at the wide end into the narrow, or sharp, end. That focused force is strong enough to cut through materials.

A screw is like a ramp wrapped around an axle. The effort force turns the axle and the thread moves the load upward.

The wheel and axle convert a linear (or straight) force into a rotational motion, and vice versa. Turning the axle winds the rope and lifts the bucket upward.

The lever (top) is a stiff bar that pivots on a point, or fulcrum. In the example above, the weights of both objects cancel each other out so the bar is balanced.

The ramp (bottom) is used to lift heavy loads by replacing a short vertical motion with a longer, sloped motion.

The lever is a force magnifier. In its simplest form, exerting a small force over a large distance at one end of the lever, which pivots on a fixed point called the fulcrum, produces a large force over a small distance at the other end. Levers abound in everyday life. A wheelbarrow, crowbar, a pair of scissors are all examples of levers.

The inclined plane, which is simply a slope or ramp, is also a machine. It makes it much easier to raise an object. The horizontal distance the object has to be moved is greater than if it were hauled vertically upward, but the force necessary to shift it is much less. The wedge is another form of the inclined plane. The thin end of the wedge pushes against an object with greater force than it takes to set the wedge in motion. As a result the wedge cuts through materials. All knives and axes are wedge machines. The screw is also an inclined plane, this time wound around a central shaft. Turning the screw moves the thread of the screw, the inclined plane, backward and forward. It takes less effort to turn a screw into an object than it would take to push it straight in (as a wedge).

The wheel and axle work together like a rotating lever. A small effort applied to the wheel turns it through a large distance, which results in a large load attached to the axle moving through a smaller distance. This is how a winch works. If the force is applied to the axle the wheel becomes a distance multiplier. The effort used to turn the axle results in the wheel traveling a much greater distance, a principle that is used to good effect on a vehicle, for example.

The pulley is a rope looped around a wheel. A one-wheel pulley simply changes the direction of the force. In a compound pulley the load is divided between two or more wheels, so a small effort can lift a large load—although a lot more rope has to be pulled to move the load a small distance.

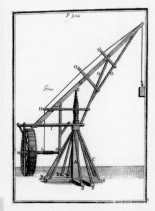

ROMAN CRANE

Roman cranes used a slave-powered pulley system to lift weights. The largest, called the polyspastos, had an impressive lifting power. The 53-ton blocks that make up Trajan's column in Rome were lifted to a height of 112 feet (34 m). The motive power to haul the rope through the pulleys came from treadwheels or treadmills turned by slaves walking inside.

22 The Pantheon

WHEN IT WAS COMPLETED, THE PANTHEON in Rome was the largest domed structure in the world and remained so for over 1,300 years. It is still the world's largest unreinforced solid concrete dome.

CONCRETE

Concrete, from the Romans on, has been a versatile building material. It is made of aggregate—grains of sand or rock—bonded together by a fluid cement. The cement is made of silicate minerals, which react with water and air to set solid, locking the aggregate together.

The building that stands today was commissioned by Emperor Hadrian and finished about 125 CE. The concrete used in the Pantheon was made using a mixture of volcanic pozzolana, lime, and a small amount of water to form a mortar. Concrete is a Roman invention. The Greeks probably used lime-based mortars but the idea of combining a mortar with an aggregate, such as rock fragments, to make concrete was begun by the Romans in the 2nd century BCE. That mixture could be molded into shape when wet and then it set rock hard.

The engineers lightened the load on the dome of the Pantheon by mixing lighter aggregates in the concrete the higher up the dome they went. Near the base heavy basalt was used, but at the top pumice, the lightest rock, was mixed in. Empty clay jugs were embedded in the upper reaches of the dome to lighten the load even further. The dome was constructed in a series of courses or steps that tapered from bottom to top, thickest at the base and thinnest at the oculus, or eye, that opens at the apex of the dome.

The Pantheon is one of the most impressive and best preserved buildings from ancient Rome. The rotunda, a massive domed building, is 142 feet (43.3 meters) in diameter, and also 142 feet in height (half of which comes from the vertical walls). The dome forms a perfect hemisphere and the 29-foot (8.8-meter) diameter oculus, the "eye" of the Pantheon, opens on to the sky above.

23 Paper

IN 105 CE HISTORICAL RECORDS SHOW THAT TS'AI LUN, an official of the Imperial Court, announced the invention of paper to the Chinese emperor.

The actual invention, however, is likely to have taken place some 200 years earlier. Fragments of paper found in the ruins of Dunhuang in China's northwest Gansu province appear to date from the reign of Emperor Wu between 140 and 86 BCE. The earliest Chinese method of papermaking most likely involved making a suspension of hemp waste in water, which was washed, soaked, and beaten to a fine pulp with a wooden mallet. A sieve of coarsely woven cloth stretched across a bamboo frame was used to collect a layer of the hemp slurry, which was then left to dry, forming a sheet of paper. Ts'ai Lun is said to have made his paper using mulberry bark, hemp waste, old rags, and fishing nets.

This Chinese edition of the Buddhist Diamond Sutra from 886 CE is the earliest example of a book printed on paper. The words and pictures were cut into woodblocks, coated with ink, and pressed on to the paper.

24 Gunpowder

THE CHINESE ALCHEMISTS WHO DISCOVERED GUNPOWDER were actually seeking a way to prolong life. For centuries potassium nitrate, or saltpeter, had been a common ingredient in elixirs that were intended to lengthen life spans.

Sometime around 850 CE, an unknown dabbler thought to add sulfur and charcoal to the mix. It was, quite literally, an explosive combination. According to a contemporary source "smoke and flames result, so that their hands and faces have been burned, and even the whole house where they were working burned down."

The first use of gunpowder in warfare was in incendiary arrows, which carried gunpowder wrapped in paper or bamboo, lit by a fuse. Later, fire lances, a sort of primitive flamethrower, and the first rockets, like today's fireworks, were developed. China retained its monopoly on gunpowder until the 13th century, but eventually the secret spread to the Arab world and from there to Europe, where it was taken up as enthusiastically as it had been by the Chinese. By 1350 the English and French armies were facing each other armed with cannons.

...rthold Schwarz, ...mythical German ...hemist, was long ...edited (wrongly) with the ...scovery of gunpowder in ...e 14th century. Schwarz ...eans "black" in German, ...d for many centuries ...npowder was known as ...lack powder."

25 The Compass

THE COMPASS FIRST APPEARED IN CHINA AROUND THE FOURTH CENTURY BCE, but it wasn't used as a navigation aid at that time. The Chinese called compasses "south pointers" and used them to tell fortunes and to ensure their homes were oriented in an auspicious direction.

These early compasses used lodestones as their pointers. Lodestone is a form of the mineral magnetite, named after the Magnesia region of Thessaly, Greece, which was an important center of iron production. The oldest reference to the properties of lodestones dates back to 600 BCE when the Greek philosopher Thales noted that they attracted iron.

There are references in a Chinese manuscript of 1040 CE, to "an iron fish" suspended in water that pointed to the south. The Chinese also discovered that an iron needle could be magnetized by rubbing it with a lodestone and that this needle would also point south.

Pointing the way

The first record of the magnetic compass in Europe is in Amalfi, Italy, at the beginning of the 14th century. But whether the compass found its way to Europe along the trade routes from China or whether it was discovered independently is a matter for speculation. Whatever the answer, it soon became of incalculable importance to the growing maritime powers, such as Portugal, England, and Spain. The compass is said to have also contributed to the rise of the Italian city-states in the 14th century. Prior to this, sailors had generally stayed within sight of land and were reluctant to venture out to sea. With ships now equipped with the means to find direction at any time, Mediterranean and then global trade grew rapidly.

A Chinese geomancy compass was used to indicate the most auspicious location for buildings according to traditional beliefs. The spoon-shaped lodestone was free to swing around on the polished plate.

WILLIAM GILBERT

William Gilbert's book *De Magnete* (*On the Magnet*), published in 1600, was an account of his researches into magnetism; it soon became the standard work on the topic. He was the first to use the terms electric attraction, electric force, and magnetic pole. Gilbert's experiments led him to conclude that the reason a compass points north–south is that the Earth itself acts as a bar magnet.

26 Galleon

UNTIL THE 15TH CENTURY, MOST SHIPS WERE BUILT IN THE STYLE OF VIKING VESSELS with sturdy hulls and a single mast. Then faster and larger ship designs appeared, culminating in the cargo vessels known as galleons.

The Viking-inspired ship was known as a cog. It was stable in rough seas but small and slow. In the 15th century, Spanish shipbuilders developed the carrack. Unlike a cog's clinker structure made of overlapping planks, it had a smooth hull made from planks fitted edge to edge, and extra masts added sails. The carrack was still slow and bulky, and the Portuguese improved the design into streamlined caravels which had triangular lateen sails. These ships, equipped with compasses, began exploring far from Europe—including to the Americas. As the risk of conflicts grew on the high seas, larger ships were built to carry more cargo and more weapons. These massive fighting ships became known as galleons—Spanish for "big ship"—and like the trireme of ancient Greece before them, they became the premier warships until the arrival of ironclads in the 19th century.

RUDDER

From the 14th century, new ships were all fitted with a rudder. Rudders had been developed over the previous thousand years, but most ships in those years were steered with a paddle on the right side. This steerboard side became known as starboard. The ship would dock with the left side against the harbor wall to prevent the steerboard from being crushed, so left on a ship is known as port.

…leons, like the earlier … designs, had "castles" …raised decks at the front … rear to give sailors a …ght advantage during …tles. In 1570, English …pwright John Hawkins …lized that a forecastle …ated drag in the wind, … so developed a sleeker …ign with less of a …perstructure over … bow.

27 Great Wall of China

THE GREAT WALL OF CHINA, OR THE WANGLI CHANGCHEN (the 10,000-Li Long Wall), is undoubtedly one of the engineering wonders of the world and one of the largest construction projects ever undertaken anywhere.

The Great Wall as it stands today was built to keep out invaders from Mongolia, who had threatened China, and the rest of Asia, many times before.

Built over a 2,000-year period, and stretching across northern China, the Great Wall is in fact many walls. Around the 7th century BCE the state of Chu began work on the Square Wall, a fortification in the north of its capital province. From the 6th to the 4th centuries other states built their own defenses. The Qi state, for example, built an extensive perimeter wall of earth and stone terminating at the Yellow Sea that incorporated mountain terrain and existing river dikes, as well as newly constructed fortifications. To protect itself from attack from the north and south, the Yan state built two separate defensive walls, the Northern Wall and the Yishui Wall. The Northern Wall was the last section of the Great Wall to be erected during the Warring States period of China's history.

Northern defense

In 221 BCE Shihuangdi, the first Qin emperor, completed the unification of China. He ordered that the fortifications set up between the warring states be removed. At the same time, he began work to link the existing wall segments in the north to form the "10,000-Li Long Wall." (A li was about 1,600 feet or 488 meters.) Hundreds of thousands of conscripted soldiers worked for a decade on the construction.

Consolidation

After Shihuangdi's death, the wall was abandoned, but during the 2nd century BCE the wall was strengthened again as part of the emperor's campaign against the peoples to the north of China. From this period the Great Wall played a role in the growth of the Silk Road trade route. In 121 BCE work was started on the Hexi Wall, or Side Wall, a project that

VISIBLE FROM SPACE

It is one of the great myths of the space age that the Great Wall can be seen from space. The idea dates back to 1754 when an English commentator felt sure the wall was so big it could be seen from the Moon! But space travelers were unable to see the wall without the aid of zoom lenses. In 2003 China's first astronaut, Yang Liwei, went into space, and confirmed to his nation that he couldn't see the structure.

HADRIAN'S WALL

Hadrian's Wall, built to protect the Roman empire's flank in what is now northern England, was one of Rome's greatest engineering projects. Work began on the 73-mile (118-kilometer) wall in 122 CE at the instigation of Emperor Hadrian and took six years to complete. Forts punctuated its 20-foot (6-meter) height at regular intervals and ditches added to its defensive capabilities. It is testament to Roman engineering that much of the wall still stands.

would last 20 years. Between the 14th and 16th centuries CE, the Ming emperors worked to strengthen the Great Wall against a possible Mongol invasion. Most of what stands today was built at the end of the 15th century. The wall was split into southern and northern lines, called the Inner and Outer Walls.

The eastern end of the Great Wall terminates at the sea at Old Dragon's Head.

Passes and forts

e structure of the wall ried from place to place incorporate natural tures. This section wall near Beijing, Chinese capital, is structed of brick. Other tions are built from stone d rock while the western tion which runs through e desert is made of mmed earth sandwiched ween wooden boards.

The wall itself was 23 to 26 feet (7 to 8 meters) high and 21 feet (6.5 meters) wide at the base, narrowing to 19 feet (5.8 meters) at the top. A low parapet along the paved top prevented traffic from falling off. Gates, or passes through the wall, were positioned at regular intervals. Guard towers above the gates, which served as command posts, were around 33 feet (10 meters) high and about 14 feet (4 meters) wide. The gates within the pass could be sealed by huge, wooden double doors. Further lines of defense such as parapets and even moats shielded the gates from attack.

28 Submarine

THE SUBMARINE HAS PLAYED A MAJOR ROLE IN NAVAL WARFARE since the First World War, but the first successful underwater excursion actually took place nearly 300 years prior to that.

A replica of Drebbel's submarine. Some reports have Drebbel releasing new supplies of breathable air from chemical reactions. Others suggested that air was supplied by snorkels floating at the surface.

In 1578, British mathematician William Bourne put forward the idea of a boat that could be submerged and rowed underwater. The boat was to be built of waterproof leather on a wooden frame and would be submerged by using hand rachets to pull in the sides, thus reducing its volume. Bourne never built his boat, and so the credit for the first actual submarine goes to Cornelius van Drebbel, a Dutch inventor. Drebbel's submarine was similar to that proposed by Bourne, with an outer hull of greased leather over a wooden frame. Oars, extending out through tight-fitting leather flaps, provided the means of propulsion. In 1620 Drebbel successfully steered his craft 13 to 16 feet (4 to 5 meters) beneath the waters of the River Thames in London, England.

DIVING BELL

Diving bells for exploring underwater were mentioned by Aristotle. His pupil, Alexander the Great is said to have had a go in one (picture). The bell works by trapping a pocket of air inside, allowing the diver to breathe. In 1689, Denis Papin attached one to bellows that could replenish the bell's air supply and a year later Edmond Halley worked out how to pressurize the air, allowing the bell to reach greater depths.

Underwater weapon

The submarine saw its first use in naval warfare during the American Revolution. David Bushnell's *Turtle* was a walnut-shaped, one-man craft built of wood reinforced with iron bands that was

PERISCOPE

The periscope is often used by submariners to see above the surface. It works by bouncing light through two mirrors or prisms, allowing the observer to see something that is not in their direct line of sight. The inventor of the periscope is unknown, though Johannes Gutenberg, of printing press fame, is said to have sold periscopes to pilgrims at a religious festival in the 1430s.

powered by hand-cranked propellers while underwater. The plan was that the *Turtle* would approach a British warship underwater and attach a gunpowder charge to its side, but none of its missions were successful.

Fulton's *Nautilus*

In 1800, while in France, American inventor Robert Fulton built the submarine *Nautilus* financed by a grant from Napoleon Bonaparte. Completed in May 1801, *Nautilus* held enough air to supply four men for three hours and submerged by taking water into ballast tanks. Two horizontal fins controlled the diving depth, and an observation dome allowed the sailors to see where they were going. *Nautilus* was intended to attach an explosive charge to the hull of an enemy ship and Fulton successfully sank a moored schooner, but his submarine proved incapable of catching British warships so France lost interest in the submarine. Fulton got Congressional backing to build a larger steam-powered craft in the United States but he died before he could complete it and work was abandoned.

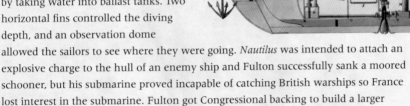

Fulton's submarine was made of copper sheets. A collapsible mast and sail were used when it was on the surface, and a hand-turned propeller when it was submerged.

29 Vacuum Pump

IN 1650, PHYSICIST AND ENGINEER OTTO VON GUERICKE INVENTED THE FIRST AIR PUMP. With it, the German showed the great force applied by the weight of the atmosphere.

…n Guericke created …uums by pumping air …ectly from the vessel, …ther than filling it with …ater first.

In his early experiments a few years previous to this, von Guericke attempted to produce a vacuum by using a suction pump to remove water from a sealed wooden cask. Not surprisingly, the cask took in air as the water was pumped out, and von Guericke started using metal containers instead and had more success. His experiments led him to conclude that a spherical shape was best at withstanding the pressure differences his pump created. Von Guericke demonstrated that light could cross a vacuum but sound could not, but he is best known for his Magdeburg experiments of 1654, in which two copper hemispheres about 14 inches (36 centimeters) in diameter were placed together, and the interior emptied of air. Two teams of horses were invited to pull them apart but failed completely, even though it was only the air around them that held the spheres together. This was an impressive demonstration of the huge power of atmospheric pressure.

30 Pendulum Clock

THE STORY IS OFTEN TOLD OF HOW GALILEO GALILEI watched a chandelier swinging in Pisa Cathedral. He timed the swings and discovered that no matter how wide the swing, it always took the same time.

Christiaan Huygens admires his pendulum clock. This clock was the first to use an oscillator to keep time. Most modern clocks now use oscillators, including digital watches, which keep time by using the vibrations of quartz crystals.

Galileo's discovery was that the swing time, or period, depends only on the length of the pendulum: No matter how heavy it is, or how wide it swings, the same length will always take the same time to go back and forth. Galileo later described how a pendulum might be used to make a clock and even produced a design for one, although he never made one. Dutch scientist Christiaan Huygens later showed that Galileo's observations were only true for swings of small amplitude. As an astronomer, Huygens was interested in accurate timekeeping. Inspired by his investigations into pendulums, he went on to invent the first successful pendulum clock in 1656. Huygens solved the problem of making the period of a pendulum truly constant by devising a pivot that caused the pendulum's weight to swing along the arc of a cycloid, not a circle. A cycloid is the flattened curve traced by a point on the circumference of a circle rolling along a straight line.

31 Steam Engine

THE STEAM ENGINE WAS THE ENERGY SOURCE that powered the Industrial Revolution. Its development spurred major advances in manufacturing, transportation, and agriculture, and fundamentally changed the world.

The first practical steam engines were developed with a particular purpose in mind: To solve the problem of flooding in mineshafts. In 1698, Thomas Savery, an English engineer, patented a machine that used steam pressure to pump water from flooded mines. Savery had studied the findings of Denis Papin, the inventor of the pressure cooker. Papin had ideas for a steam engine powered by a cylinder and piston, inspired by his pressure cooker observations, but they hadn't proved practicable.

Savery's machine consisted of a boiler, a water-filled reservoir, and a series of valves. Steam was fed into the reservoir, pumping the water out through a valve until the reservoir was empty. Coolant water was then sprayed on the reservoir, causing the steam inside to condense, thus creating a vacuum that drew up more water through a second valve. Slowly and not that surely—the cast iron components often split—

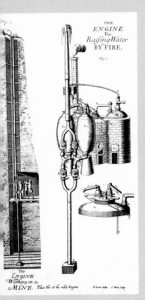

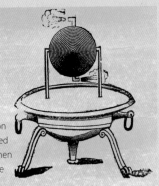

water could be raised from the flooded mine, although there was a limit to how high the machine could lift water.

Newcomer Newcomen

In 1711, Thomas Newcomen developed a system using a redesigned steam engine. Newcomen's steam pump consisted of a cylinder fitted with a piston—another design inspired by Papin's earlier ideas. When the cylinder was filled with steam, a counterweight moved the piston to the extreme upper end of the stroke. As in Savery's engine, coolant water was used to condense the steam in the cylinder, creating a vacuum. Atmospheric pressure acting on the piston pushed it back down in the cylinder, and the resulting force raised the plunger in a water pump.

In 1712, Newcomen built the first piston-operated steam pump for the Conygree Coalworks in central England. Newcomen's "atmospheric" engine was not without its drawbacks. For one thing, it was highly inefficient, converting only about one percent of the thermal energy in the steam to mechanical energy. Nevertheless, it remained the best engine for the next 50 years.

FIRST STEAM ENGINE?

The aeolipile, invented by Heron of Alexandria in the first century CE, was perhaps the first steam turbine, transforming the energy of steam into motion. It was simply a hollow sphere with a pair of tubes that fed steam into the sphere from a cauldron below. Two more bent tubes projected from the centerline of the sphere. When the steam escaped the resultant force caused the sphere to revolve.

Savery's steam engine was called "The Miner's Friend; ..., An Engine to Raise ...ater by Fire."

Watt's engine

In 1765, Scottish engineer James Watt modified a Newcomen engine by adding a separate condenser that made it unnecessary to cool the cylinder after heating it. Keeping the cylinder and piston at steam temperature while the engine was operating brought fuel costs down by about 75 percent. In addition, Watt obtained increased power by applying steam to both sides of the piston.

Watt's rotative engine enabled the steam engine to be used to operate all kinds of machines in factories and cotton mills. The rotative engine was widely adopted, and it is estimated that by 1800 Watt and his business partner Matthew Boulton had built 500 engines, most of which were of this type.

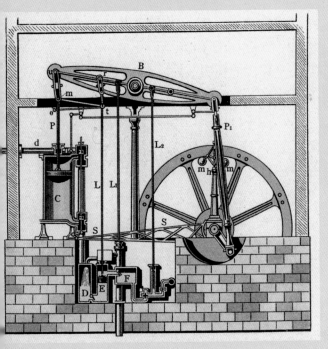

Watt's breakthrough engine had an arrangement of gears that converted the up-and-down movement of a piston into circular motion. A heavy flywheel evened out the variations in the force delivered to the engine shaft by the action of the piston in the cylinder, and a governor connected to the flywheel regulated the flow of steam to the engine.

32 Seed Drill

UNTIL THE 18TH CENTURY, FARMERS sowed seeds by broadcasting—walking while throwing handfuls of seed randomly onto plowed land. This was wasteful as some seeds grew too close together and others failed to grow.

Agriculturalist Jethro Tull determined to do something about this inefficiency. In 1701, he developed a horse-drawn, mechanical seed drill—so-named to reflect the method of making a hole for a seed. Tull's drill had a hopper to hold the seeds, which dropped on to a rotating grooved cylinder that fed the seed to a funnel below. The seeds were thus directed into a channel dug by a plough at the front of the machine, and were immediately covered by a harrow attached to the rear. Planting the seeds at regular intervals, at a consistent depth, and in a straight line limited waste, dramatically increased yields, and made it much easier to gather in the harvest.

Tull's seed drill allowed farmers to sow three rows of seeds simultaneously.

33 Balloon

LIGHTER-THAN-AIR FLIGHT IN BALLOONS WAS THE FIRST MEANS by which people took to the skies. The age of flight began in 1783.

The principle of buoyancy has been understood since Archimedes figured it out in the 3rd century BCE. Any object placed in a fluid (gas or liquid) is acted upon by an upward, or buoyant, force equal to the weight of the fluid displaced by the object. If a balloon can be made lighter than air, it will be pushed up and float away. One way of doing this is to heat the air inside a balloon, making it less dense (and weigh less) than the air around it. Another method is to fill the envelope with a "lighter-than-air" gas, such as hydrogen or helium.

French paper manufacturers Joseph-Michel and Jacques-Étienne Montgolfier began experimenting with lighter-than-air devices after observing that heated air made a paper bag rise. They made themselves a 33-foot (10-meter) wide balloon and launched it 6,500

HYDROGEN BALLOON

In December 1783, Frenchman Jacques Alexandre César Charles and a companion went up in a hydrogen-filled balloon that flew longer and higher than the Montgolfier brothers' hot-air balloon, also demonstrated that year. Charles had actually launched his first hydrogen balloon in August. When it came back to Earth it was attacked by terrified peasants who thought it a monster from the sky.

The year 1783 saw the inaugural flights of hot-air balloons and hydrogen-filled balloons. Both were launched from France—though not at the same time or in the same place as our picture suggests. Hot air at the sort of temperatures used in hot-air balloons weighs about four-fifths as much as cold air. Hydrogen weighs around one-fourteenth as much.

feet (1,980 meters) into the air on 4 June 1783. Next the Montgolfiers decided to send up a sheep, a duck, and a rooster in a basket beneath the balloon. The crew selection was not random: The duck was assumed to be unaffected by altitude, while the rooster was a bird but one that could not fly high. The sheep was the stand-in for a human aviator. All three were unaffected by the eight-minute flight witnessed by a crowd of 130,000, including the French king. The next step was to get a person aloft and the following month, on 15 October, Jean-François Pilâtre de Rozier ascended in a tethered Montgolfier balloon, becoming the first person recorded to fly.

34 Hydraulics

THE ENGLISH LOCKSMITH AND PROLIFIC INVENTOR JOSEPH BRAMAH
patented the first hydraulic press in 1795. All hydraulic
machinery, from excavators to robot arms, work the same way.

Bramah's press had two cylinders and pistons with heads of different areas. Like a lever, the hydraulic press acts as a force multiplier. A force exerted on the smaller piston is translated into a larger force on the larger piston. The difference in the two forces is proportional to the difference in surface area of the two pistons. The distance over which the force is applied to the smaller piston is greater than that traveled by the larger piston, in exactly the same ratio as the force is multiplied. So if the force is doubled, the larger piston will only travel half the distance. It's possible to create huge force using hydraulics. For example, if the second piston is a hundred times bigger than the first, the force is multiplied a hundred times. But of course in this case the second piston only moves a hundredth the distance of the first.

The hydraulic press uses Pascal's Principle, which states that if you apply a force to part of an enclosed fluid, that force is transmitted to every part of the fluid. Therefore, pushing on the lever forces the fluid (green) through the system and results in a push on the piston.

35 Industrial Revolution

AT THE START OF THE 19TH CENTURY, A FUNDAMENTAL SHIFT BEGUN, at first in Britain and later around the world. Rural societies were transforming to primarily urban and industrial ones.

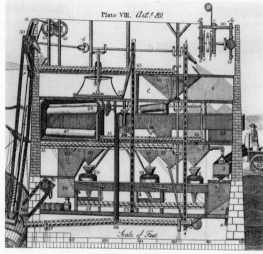

In the 1790s, Oliver Evans, an American engineer, developed an automated mill. These mills were sponsored by the new United States government to boost the quality of flour produced in the United States after its supplies from Britain had ended after independence.

The growth of the textile industry was a prime mover of the Industrial Revolution. Previously, merchants would provide raw materials and basic equipment to people working in their own homes (so-called cottage industries) and then later collect the finished product. This wasn't an efficient way of working. In the 1700s, a series of innovations led to ever-increasing productivity in textile manufacture, while at the same time requiring less human involvement. Around 1764, Englishman James Hargreaves invented the spinning jenny ("jenny" is though to be a diminutive of "engine"). This allowed one person to produce multiple spools of threads simultaneously. Other inventions followed, such as the power loom, developed by Edmund Cartwright in the 1780s, which mechanized cloth-weaving. It made sense for these labor-saving devices to be used in large numbers inside a purpose-built manufactory—or factory.

Hot metal

Developments in iron and steam also gathered pace. Steam helped open up the mines that supplied the coal that powered the furnaces that smelted the iron. Iron and steel were essential components in the building of the new machines. In the early 18th century, Abraham Darby discovered a cheaper, easier method of producing cast iron, using coke rather than charcoal to fuel his furnaces. In the 1850s, Henry Bessemer developed an inexpensive process for mass-producing steel. Steam and steel came together in locomotives and ships that transported goods.

This steelworks in eastern Germany was set up in 1868 and was one of the first in that country. In the last quarter of the 19th century, Germany industrialized very fast.

POWER SUPPLY

One of the reasons the Industrial Revolution took hold in Britain was the abundance of coal in that country. Coal production there soared from 5 million tons in 1750 to 50 million tons in 1850. As coal powered steam, so steam drove mining. Steam-powered pumping engines allowed shafts to be sunk deeper. Nevertheless, technology did not improve the lot of the miners. Accidents were common, and boys were sent down the pits along with the men.

Roads and rails

Transportation was another part of life that was transformed by the changes brought about by the Industrial Revolution. Before the harnessing of steam power, raw materials and finished goods were moved by horse-drawn wagons on roads that were often poorly maintained, on rivers and canals, and by sailing ship from one country to another. In 1803, British engineer Richard Trevithick built the first railroad steam locomotive. By 1830, the Liverpool and Manchester Railway became the first to offer regular steam passenger services, and by 1850, Britain had nearly 6,300 miles (10,140 kilometers) of railroad track. In the 1820s, Scottish engineer John McAdam developed his "macadamization" process for road construction, where roads were dressed in gravel, making them smoother and more durable.

The iron bridge, built over the River Severn in Shropshire, England in 1779, was the first of its kind in the world. The location is now a village named Ironbridge.

Worldwide movement

Industrialization had its downside as skilled craftspeople were replaced by machines, and workers flocked from the countryside into the sprawling urban areas that grew up around the factories. Housing became overcrowded and inadequate. From about 1760 to 1830, the Industrial Revolution was largely a British phenomenon. In an attempt to keep hold of their advantages the British blocked the export of machinery and skilled workers to other countries. But Britons seeking profit overseas found ways to take their know-how abroad. William and John Cockerill set up machine shops in Belgium, which became the next country to industrialize, developing its iron, coal, and textile industries. France, busy with a revolution of its own, lagged behind, but by the mid-19th century it too was an industrial power. Germany, although rich in natural resources, did not begin to exploit its potential until it became a unified country in 1870. Across the Atlantic, American industrial power began to outstrip that of Europe as the 20th century approached.

LUDDITES

Today, someone who doesn't like technology is called a "luddite." That name goes back to a 19th-century labor movement in England, whose members smashed up textile machinery when their demands for better conditions were refused. They took their name from Ned Ludd, a mythical figure who was said to have started the movement. The disturbances carried on sporadically for years until heavy reprisals by the government, including execution and transportation, brought them to an end.

36 Electric Motor

IN 1821, MICHAEL FARADAY, AN ENGLISH SCIENTIST, INVENTED ONE OF THE MOST UBIQUITOUS devices in the modern world—the electric motor. His invention built on discoveries from a new field of physics: Electromagnetism.

In 1800, Alessandro Volta had invented the electric cell, a reliable source of electric power; within weeks scientists were using it in a variety of experiments. The first hint that electricity could produce movement came in 1820 when Danish scientist Hans Christian Øersted noticed that the needle on a compass moved when it was brought near a wire carrying an electric current: Electricity produced magnetic effects.

The original electric motor presented by Michael Faraday in 1821.

The Saxton Magneto Electric Machine, invented in 1833 by American Joseph Saxton, was an early dynamo. Turning the handle generated an electric current.

In a spin

In 1821, Michael Faraday demonstrated that a suspended wire carrying current would circle around a magnet. This was the first electric motor, although it had no means of transmitting its motion to machinery. In 1831, Faraday and Joseph Henry, working independently in America, discovered that Øersted's findings worked in reverse. Not only did a moving electric current produce magnetism, a moving magnet produced an electric current. This was the principle of electromagnetic induction and laid the foundations for the invention of the electric generator, which is a motor in reverse.

A contender for the first real electric motor, in the sense that it could actually do work, was built by Moritz Hermann von Jacobi in Prussia 1834. Using a horseshoe electromagnet he constructed an electric motor that generated about 15 watts of power. Another version was produced by Thomas Davenport, a blacksmith from Vermont. Using equipment that he bought from Joseph Henry, Davenport also built an electric motor in 1834. In 1837, he used his motor to run a small car, possibly the world's first electric car. Davenport found new uses for his motors, including silk weaving, operating a lathe, and running a printing press, which he used to publish a book: *The Electro-Magnetic and Mechanics Intelligencer*.

37 Water Treatment

AN ADEQUATE SUPPLY OF FRESH, CLEAN WATER IS ABSOLUTELY VITAL FOR HEALTH. Methods of purifying water go back over 4,000 years, well before any knowledge of bacteria or other pathogens.

Sanskrit writings have been found advocating boiling and filtering water through sand and charcoal. The first citywide water filtration plant was designed by Robert Thom in Paisley, Scotland in 1804. Thom used slow sand filters, which trapped contaminants. In 1827, James Simpson created a similar design to Thom's which he was soon implementing in municipal water-treatment plants throughout England.

In mid-19th century London, diseases such as cholera spread quickly through a tightly packed population forced to rely on poor quality drinking water. It was noted that in areas where sand-water filters had been installed, the outbreaks of cholera had greatly decreased. Based on this evidence, the Metropolis Water Act of 1852 was passed, the first legislation of its kind, and sand-water filters were installed throughout the city.

London in the mid-19th century was effectively a city that had outgrown its infrastructure. It was producing huge amounts of sewage that were being tipped straight into the River Thames, which was little more than an open sewer. The hot summer of 1858 created the "Great Stink of London," a miasma so foul it brought Parliament to a halt. In response Joseph Bazalgette, chief engineer of the city of London, built a series of low-level sewers that directed the flow of foul water to new treatment works. The low-level sewers were built behind riverside embankments, reclaiming the muddy banks of the river, and creating land that could be used for roads and public gardens.

THOMAS CRAPPER

Thomas Crapper was a highly successful manufacturer of flush toilets in the 19th century, but he did not invent them. The first flush toilet was described by Sir John Harington in 1596. It had an oval bowl waterproofed with pitch and resin that was flushed by 9 gallons (34 liters) of water from an upstairs cistern. Harrington remarked that when water was scarce, up to 20 people could use his commode between flushes!

Cartoons highlight the terrible condition of Old Father Thames and his rejuvenation following the installations of sewers.

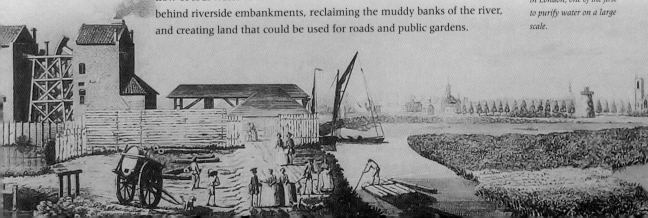

The waterworks at Chelsea in London, one of the first to purify water on a large scale.

38 Refrigeration

FOR THOUSANDS OF YEARS THE ONLY WAY PEOPLE had of cooling their food was ice. But ice wasn't readily available to everyone and, unless you built yourself an icehouse, it couldn't be kept throughout the year.

One of the first breakthroughs in building a refrigerating machine was made by Scottish chemist William Cullen in 1748. He used a vaccum pump to make a liquid chemical boil. As it did so it sucked heat from the air, creating a temperature drop that made water around it freeze. In 1805, American Oliver Evans proposed a cooling system that used this effect by evaporating ether and then condensing it again in a closed system. Evans never built his machine but later engineers had a go. Fellow American Jacob Perkins won the first patent for a refrigerator, but his machine was too inefficient to compete with the natural ice industry. In 1851, the first practical refrigerator and ice-making machine was made by James Harrison, in Geelong, Australia.

James Harrison's ice-making machine was ma commercially viable beca there were few sources of natural ice available in sultry Australia.

39 Ocean Liner

BY 1818, STEAMSHIPS DRIVEN BY STERN- OR SIDE-MOUNTED PADDLE WHEELS were plying the Mediterranean Sea and the rivers of Germany. The race was on to make ships powerful enough to cross an ocean.

The SS Great Western powers through the Atlantic, setting the record for the ocean crossing at less than 14 days in 1838.

At first a steam engine was regarded as a backup source of power for sail. In May 1819, the SS *Savannah*, equipped with collapsible paddle wheels, made the first partially steam-powered crossing of the Atlantic. The trip lasted 633 hours (just over 26 days), steam being used for just 80 of them, but the possibilities were obvious. By 1833, steam-powered vessels were traveling to India, South Africa, and Australia. Major shipping lines competed to build huge vessels capable of crossing the Atlantic at high speed entirely under steam.

The first attempt at a transatlantic service was in 1838 by the SS *Sirius*, a small, London-based ship carrying 40 paying passengers. She sailed from Queenstown in Ireland on 4 April. Four days later, her larger rival, the SS *Great Western*, sailed from Bristol, England. The *Sirius* arrived first but had to burn its furniture after running out of coal,

…mbard Kingdom Brunel, …e pioneer of ocean-going …amships, is pictured … front of the immense …chor chains of the …eat Eastern.

…e RMS Lusitania, *the …test ship of its day, docks … New York after its first …yage in 1907.*

while the *Great Western* carried enough fuel for the full journey.

The designer of the *Great Western*, Isambard Kingdom Brunel decided he would outdo all of his rivals. In July 1839, the keel was laid in Bristol for a 3,270-ton iron ship, to be called the *Mammoth*. Eschewing convention, Brunel designed the ship to be screw-driven, the first large ship of its kind not to have paddle wheels. When completed the ship, now renamed SS *Great Britain*, had lavish accommodation for 360 passengers.

Bigger and better

Around the same time, Samuel Cunard's first ship, the *Britannia*, went into service, carrying mail across the Atlantic. She docked in Halifax, Nova Scotia, on July 17, 1840, having made the crossing in 11 days and 4 hours. The Cunard ships were fast and reliable, although they couldn't challenge the *Great Britain* for passenger comfort. By 1853, the *Great Britain*, now with accommodation for up to 630 passengers, was operating a service between London and Australia.

In 1854, Brunel began work on the SS *Great Eastern*. Nothing like this had ever been contemplated before. Carrying 4,000 passengers and enough coal to reach Australia without refueling, the *Great Eastern* was the largest ship the world had ever seen. The 692-foot (211-meter) long ship had six masts for sails as well as paddle wheels and screw propulsion. She weighed over 30,000 tons fully loaded. Not until the generation of superliners such as the RMS *Lusitania* in 1907 would the *Great Eastern* be surpassed. Sadly, the ship was bedeviled by technical problems and she never really met her potential. Brunel died just four days after the *Great Eastern* began its sea trials.

BANANA BOATS

At the beginning of the 20th century large fruit companies built refrigerated ships to carry easily perishable bananas from the tropics to the markets of America and Europe. Many of these ships had luxury accommodation for paying passengers as well, beginning the cruise industry. The fruit ships were white-washed to reflect sunlight and keep their cargo cool. Cruise ships have been white ever since!

40 Bicycle

KIRKPATRICK MACMILLAN, A BLACKSMITH FROM RURAL SCOTLAND, used his metalworking skills to build the first pedal cycle. He used it to travel around the lanes of his native Dumfriesshire but won neither fame nor fortune from his invention.

Until 1839, the quickest means of getting around was by horse, or failing that a dandy horse. This wooden-framed machine had two in-line wheels with iron tires and was propelled by pushing yourself along with your feet. The rider used handlebars to steer, and a brake to stop it. Riding downhill was fine, but traveling on flat surfaces was hard work, and it was easier to walk uphill.

Macmillan was a practical young man. As a boy he had seen someone riding a dandy horse through his village and he decided to build one for himself. This mission accomplished, he imagined how much quicker he could travel if he was able to propel the machine. Macmillan set to work on a revolutionary design and in 1839 his pedal cycle was ready to take out on the road. It was propelled by a vertical reciprocating movement of his feet on two pedals. This movement was transmitted by connecting rods to cranks on the rear wheel, causing the rear wheel to turn and driving the cycle forward, the same basic design that has been used ever since. Kirkpatrick's bicycle was heavy so the physical effort of cycling up hills was enormous. This did not deter Kirkpatrick, who was clearly very fit (although local people called him "Daft Pate")— he could travel to the town of Dumfries, 14 miles (22.5 km) away, in less than an hour.

Kirkpatrick didn't patent his invention, and his idea was developed by others. In the 1860s, French engineers built similar machines with pedals on the front wheel, known as velocipedes. In the 1880s, the first chain-driven bikes were introduced. They were known as "safety bicycles" because the chain powered the back wheel, leaving the front wheel free for steering alone.

The dandy horse, so-named because it was a plaything of wealthy young men, or dandies, was invented in Germany in 1817.

A selection of velocipede designs from the late 19th century. A large front wheel made the devices much faster. Only the chain-powered "safety" model had wheels of similar sizes.

41 The Thames Tunnel

AT THE START OF THE 19TH CENTURY THERE WAS A PRESSING NEED to connect the docks on the north and south banks of the River Thames in London. Large ships passed along this stretch of the river so it couldn't be bridged. A tunnel seemed the only option.

Digging through the soft clay of the Thames riverbed proved a problem—all early attempts simply collapsed. A tunnel under the Thames seemed impossible.

Marc Brunel, an Anglo-French engineer, refused to accept this verdict. With Thomas Cochrane he invented a protective shield that was to make tunneling in clay possible. His inspiration is thought to have been the humble shipworm, whose head is protected by a hard shell as it bores through ships' timbers. The reinforced cast-iron tunneling shield was divided into 36 compartments that protected miners as they dug at the tunnel face. As the digging advanced, the shield was driven forward by jacks and the surface behind it lined with bricks. Work began on a tunnel in 1825.

Even so, the underground workers faced plenty of other hazards. Filthy water from the river above seeped in, causing many to fall ill. Concentrations of methane regularly caused fires and flooding was a frequent hazard. Soon Marc Brunel's son, 20-year-old Isambard, joined the dwindling workforce.

The Thames Tunnel was finally completed 18 years after work started. It measured 35 feet (11 meters) wide by 20 feet (6 meters) high and ran for 1,300 feet (396 meters) at a depth of 75 feet (23 meters). Since groundwater continually seeped into it, drainage pumps were installed. Lighting, roadways, and spiral staircases were fitted and on 25 March 1843 it was opened to the public as a pedestrian walkway.

Miners dig inside a tunneling shield. The key point of the shield was that it supported the unlined ground in front and around it in order to prevent collapse.

[Th]e Thames Tunnel, [co]mpleted in 1843, was the [wo]rld's first tunnel dug [un]der a navigable river. [At] first it carried wagons [an]d pedestrians. Today, it [is] used as part of London's [rai]lroad network.

42 Steelmaking

IN THE MID-19TH CENTURY, LARGE CONSTRUCTION PROJECTS HAD TO USE CAST IRON OR WROUGHT IRON but there were problems with both materials. Steel was superior—but very expensive. All that changed in 1855.

The man responsible was Henry Bessemer, the son of a French engineer living in Britain, who showed a flair for invention from a young age. The project that set him on a course for greatness came in the 1850s during the Crimean War. The cannonballs used then were lethal if they made a direct hit but very inaccurate. Bessemer designed a long, thin projectile with spiral grooves around it (we now call this rifling) to make it spin and so increase its accuracy. The problem was that the gun barrels of the time weren't strong enough to fire these projectiles. They often shattered during firing. Bessemer needed cannon barrels made of steel, a strong alloy of iron and carbon. Cast steel was ideal but could only be made in small batches

Henry Bessemer's mass production of steel meant prices fell dramatically. It is no exaggeration to say that it accelerated the Industrial Revolution.

and was therefore very expensive. Bessemer experimented by blowing air over molten pig iron in a small furnace and discovered that he could convert it to steel. The carbon impurities oxidized, and the resulting carbon monoxide burned off. Next, he needed a design that would expose all the molten pig iron to air as quickly as possible.

ALUMINUM

Despite being the most abundant metal in Earth's crust, aluminum is difficult to refine, so much so that it was considered a precious metal. An aluminum bar was once part of the French crown jewels. The big breakthrough in aluminum production came in 1886 when Americans Charles and Julia Hall and the Frenchman Paul Héroult independently invented a method of purifying aluminum with electricity.

In 1855, Bessemer revealed his new process. Molten pig iron was poured into the top of a large pear-shaped vessel. Cool compressed air was injected into the vessel through the base of the container. Within minutes the carbon, manganese, and silicon impurities in the iron were oxidized, the carbon monoxide igniting and burning off and the other impurities forming slag, which floated to the top of the liquid metal. Critics argued that the cool air would solidify the iron prematurely, but Henry's experiments proved that the oxidations produced sufficient heat to keep the metal molten, removing the need for further expensive fuel to maintain it in the molten state.

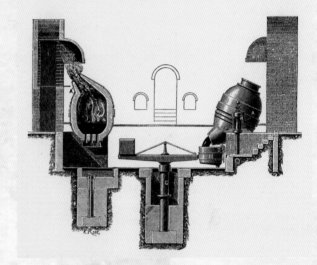

ssemer's steelmaker, known as a converter, was designed for maximum efficiency. It was upright when the air was burning off the impurities, but tipped up for the molten steel to flow out.

Where previously a 12-hour shift had produced 1.5 tons of wrought iron, ten times that amount of steel could now be made in 20 minutes in a Bessemer converter. Henry Bessemer's invention was the most important development in the steel industry, making it possible to make all kinds of objects, from railroad tracks to building frames, from superstrong steel.

43 Plastic

LIGHTWEIGHT, MOLDABLE, AND EITHER TOUGH OR DELICATE, plastics are now crucial components of pretty much everything from cell phones and automobiles to robots and replacement body parts.

collection of Parkesine objects, including a billiard ball, tube, and jewelry box.

The first man-made plastic was patented by Birmingham metallurgist and inventor Alexander Parkes in 1856. It was a nitrocellulose compound—cellulose treated with nitric acid and a solvent—which he called Parkesine. Parkes set up a factory in London, but his product quickly disappeared from public use because it was expensive, prone to cracking, and highly flammable. An improved version, Xylonite, was invented by fellow Briton Daniel Spill, but he became embroiled in patent disputes and went bankrupt. And so it was left to the American John Wesley Hyatt to produce the first successful plastics. After acquiring Parkes' patent in 1869, Hyatt experimented with Parkesine in a quest to find a replacement material for ivory billiard balls. He succeeded, inventing Celluloid, which entered mass production in 1872. His company produced, among other things, Celluloid piano keys, false teeth, and billiard balls.

44 Ironclad

ARMORED WITH A METAL-SKINNED HULL, powered by steam, and armed with guns that fired explosive shells, ironclads transformed naval warfare after they burst onto the scene in the early stages of the American Civil War.

On 9 March 1862, the world's first clash between ironclad warships took place at the Battle of Hampton Roads, Virginia, with the Confederacy's CSS *Virginia* pitted against the Union's USS *Monitor*. The *Virginia* was an old frigate (the *Merrimack*) that had been substantially rebuilt. The main decks were designed to remain underwater and were covered with 4-inch (10-centimeter) thick iron plate. An armored enclosure was built on top, with two 2-inch (5-centimeter) layers of iron plating angled at 36 degrees to the horizontal to deflect enemy fire. There were 14 gun ports, equipped with powerful guns.

CSS *Virginia* had already sunk two Union warships before the epic encounter with the *Monitor*. *Monitor*'s design was truly revolutionary. Smaller than *Virginia* but faster and more nimble, she was armed with a rotating two-gun turret. The battle went on for several hours, with both vessels taking direct hits but none of them proving to be devastating. *Monitor* eventually withdrew to assess the damage, but the battle was inconclusive. Although neither warship survived the war, their ability to withstand direct hits by explosive shells had convinced both navies to build more. In fact, the Union built 50, modeled on the *Monitor*.

The idea of metal cladding on the hulls of warships was not new. The British Navy had fitted copper plates to its warships in the 18th century but that was to protect these so-called "copper bottoms" from shipworms, not enemy shells. With the increased firepower of naval artillery in the mid-19th century, the defensive strength of iron armor was vital, and was made feasible by steam engines that were more powerful than sails, able to propel heavier, iron plated ships. The first oceangoing ironclad, the *Gloire*, was launched by France in 1859, and the British responded two years later by launching two Warrior-class ironclads: HMS *Warrior* and HMS *Black Prince*.

The first ironclad battle was an attempt by the Confederates to break a Union blockade of Chesapeake Bay. The battle had no winner, as would be the case in many engagements between heavily armored battleships in future.

45 Subway

WHEN AN ENGLISH LAWYER SUGGESTED building an underground railroad to ease traffic jams on busy streets, people at first didn't take him seriously. But his plan did eventually transform the way people travel around cities.

After the great railroad stations of Paddington, Euston, and King's Cross were completed close to each other north of London in the 1850s, congestion in the area became intolerable. Charles Pearson's idea was an underground train between the stations to remove traffic from the street above. At first, his ideas were considered far-fetched, but in 1854 it was agreed to build the world's first subway. Just nine years later, in 1863, the Metropolitan Railway was built beneath London's streets, though sadly Pearson didn't live to see it.

Part of the subway, called the London Underground, was built by the cut-and-cover method. A deep trench was dug, rails laid, then it was covered over so that life could continue as normal above it. Other sections, dubbed "the tube," were tunneled, using the shield system developed by the Brunels. The 3.7-mile (6-kilometer) subway was an instant success, with 38,000 passengers traveling in its gaslit carriages pulled by steam locomotives on the first day. The London subway network expanded rapidly in the next few years. The authorities in other cities realized that rapid transit subway networks could ease many of their street congestion problems. In 1896, both Budapest in Hungary and the Scottish city of Glasgow opened their own underground rapid transit networks. The first line on New York City's Metro carried passengers in 1904, and the Moscow Metro followed suit in 1935.

MAPS

Harry Beck was a technical draftsman familiar with drawing electrical circuit diagrams. He believed that London subway passengers were more concerned with which line went to their destination and which interchange they needed than they were with geographical accuracy. So he designed a diagrammatic map with the lines shown in different colors. When the map was published in 1933 it was an instant success. With a few modifications it is still used today, and the concept has been copied worldwide

A Metropolitan Railway train passes Praed Street junction under the streets of Paddington, London.

46 The Transcontinental Railroad

BEFORE 1869, A JOURNEY EAST TO WEST across the United States took six months by wagon. After that year, the journey took just over a week, thanks to a railroad running all the way to the West Coast.

The Transcontinental Railroad was a triumph of engineering, conquering steep-sided ravines and precipitous mountains with hundreds of bridges and dozens of tunnels. The railroad revolutionized the economy and settlement of the American West, making transportation of goods and people faster, cheaper, and more flexible. The 1,907-mile (3,069-km) "Overland Route" ran from Council Bluffs, on the eastern bank of the Missouri River, to Sacramento on the California coast.

Seven years previously, the Pacific Railroad Act had chartered two railroad companies to build it: The Central Pacific (CP) and Union Pacific (UP). Each were paid in land and a fee for every mile of track laid, so the project became a race. The Central Pacific, starting from the west, faced the greatest problems. Because California was isolated, building materials—everything from picks and shovels to rails and

Building the railroad was hard and dangerous work. It is estimated that more than 1,000 workers died during the six years of construction.

GOING LOCO

Early trains crossing the American continent were hauled by wood-fired locomotives. They were slow and struggled with steep gradients. To heave Central Pacific trains over the Sierra Nevada a very different type of loco was needed from those that bowled across the flat deserts of Nevada. Early wood-fired locomotives, using the plentiful supplies of wood from the eastern forests and foothills of the Rockies were gradually replaced after the 1870s by bigger and bigger coal-fired engines.

Cow-catcher cleared the track

Conical smokestack prevented sparks escaping

Fuel carried in tender

locomotives—had to be brought there by ship. Most cargo went 18,000 miles (29,000 km) around Cape Horn, at the southern tip of South America, in some 200 days. A route using the new Panama Rail Road was faster (40 days) but twice as expensive.

Climbing to Donner Pass in the Sierra Nevada Mountains, the rail track had to rise very steeply—7,000 feet (2,133 m) in 90 miles (145 km). In the mountains, CP's teams of mostly Chinese laborers had to cut 15 tunnels, including the 1,659-foot (506 m) Summit Tunnel. The rate of advance was painfully slow: Just one foot a day until nitroglycerine was introduced to blast the rock. Snow was another hazard in the mountains. To keep the higher sections of new track open in winter, 37 miles (59.5 km) of wooden snow sheds were built overhead to protect the line from avalanches. Once out of the Sierra Nevada, the rate of tracklaying accelerated. On one occasion, a CP crew laid 10 miles (16 km) of track in a day—a record that still stands today.

Advance from the east

Union Pacific started from the eastern end and initially passed over the relatively easy terrain of the Great Plains before running into the foothills of the Rockies. However, its engineers and tracklayers faced an additional problem: Angry Native Americans who saw their land being taken from them.

Teams went ahead to build wooden trestle bridges over ravines. One, the Dale Creek Bridge in Wyoming, was 650 feet (198 m) long and rose 125 feet (38 m) above the valley bottom. Innovative construction techniques were needed, with the bridge components prebuilt of timber in Chicago and shipped on rail cars to the site for assembly. When the CP and UP teams met in May 1869 they had laid 690 miles (1,110 km) and 1,085 miles (1,746 km) of track, respectively, in just over six years. The United States was now properly united.

GOING INTERCITY

The Liverpool to Manchester Railway, designed by George Stephenson and opened in 1830, was the world's first railway that connected two cities. Industrialists in the two cities needed to get raw materials from the port of Liverpool to the factories of Manchester, and finished goods back to the port. They had previously used the Bridgewater Canal but now wanted a quicker and cheaper means of moving goods. The railway carried passengers as well as goods.

On May 10, 1869, at Promontory Summit in Utah, rail entrepreneur Leland Stanford hammered in a golden spike that joined the last rails of the First Transcontinental Railroad, linking the east coast of the United States with the west.

47 Telephone

THE SAME BASIC PRINCIPLES EXIST FOR MODERN CELL PHONES as they did for the very first telephones. For the early pioneers the big problem was how to convert sound waves into electric signals.

A telephone consists of a microphone to speak into, a device to change your voice into electric signals, a means of sending these signals to their destination, and a loudspeaker at the destination to change the electric signals back into your voice.

Scottish-Canadian Alexander Graham Bell is generally credited with inventing the telephone to transmit the human voice. Bell's mother had gone deaf, and his father developed a way of teaching deaf people to speak, a method which Alexander taught. He was also intrigued by the challenge of helping the deaf to hear. In 1874, he built a device which he called a phonautograph. By speaking into a dead man's ear, he made the ear's membrane vibrate. The vibrations caused a lever attached to the ear to produce a wave pattern on smoked glass. The louder the sound, the bigger the wave. For the next two years he worked on ways to convert sound waves into electric current rather than the movements of a lever.

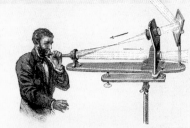

PHOTOPHONE

Bell's photophone was the world's first wireless telephone system. Words spoken into a tube made a mirror vibrate, changing the amount of sunlight reflected back to a receiving mirror, where light waves were converted back to sound. There were practical problems—it didn't work on cloudy days, for example—and it was never used widely. However, it's now recognized as the forerunner of fiber optic telecommunications used to transmit telephone and internet signals at light speed.

Liquid transmitter

Bell's big breakthrough came on March 10, 1876, when he publicly unveiled his "liquid transmitter" for the first time. This ingenious device consisted of a vertical cone with parchment stretched like an eardrum across a hole at the bottom. A cork with a needle attached was fixed to the outside of the parchment and this dipped into a small container of dilute sulfuric acid. The needle was wired to a battery. When Bell spoke into the open top of the cone, the sound waves made the parchment and the needle beneath it vibrate. The vibrations varied the current passing between electrical contacts, so sound waves were converted into electric signals. The signals passed along a wire to the receiver, which, during the successful experiment, was in the charge of Thomas Watson in another room. There, the electrical impulses were

The Reis "telephon" from 1861 was the first device to send a voice signal by a wire, although it only worked in one direction.

I

B

II

A

Alexander Graham Bell at the opening of the long-distance line from New York to Chicago in 1892.

changed into sound. While Bell was setting up the experiment he spilled some acid on his trousers and spoke down the line to his assistant, "Mr Watson, come here. I want to see you." Within months Bell was transmitting voice messages over much longer distances.

Horses and cucumber salad
Fifteen years before Bell's triumph, the German inventor Philipp Reis had built a prototype telephone that worked in a similar way, with sound converted to electricity and back to sound again. In his first receiver he wound a coil of wire around an iron knitting needle and rested the needle against the F hole of a violin. As current passed through the needle, the iron shrank a little, and a click was produced. Reis coined the term "telephon" to describe his device.

Thomas Edison later credited Reis and Bell for their achievements, writing: "The first inventor of a telephone was Phillip Reis of Germany, only musical not articulating. The first person to publicly exhibit a telephone for transmission of articulate speech was A. G. Bell." This didn't fully do justice to Reis's achievement, because he had managed to transmit speech—the peculiar phrase, "Das Pferd frisst keinen Gurkensalat," meaning "The horse does not eat cucumber salad," chosen because those words are difficult to understand acoustically in German.

Bell's device was indeed the first practical commercial telephone for transmission of articulate speech—although American Elisha Gray is also noted for independently developing a telephone at almost exactly the same time. Edison improved on Bell's microphone by using carbon, which changes electrical conductivity under small amounts of pressure. The human voice could now be carried over much longer distances, and more clearly. Carbon-based telephone microphones were still being manufactured in the 1970s.

CROSSED LINES
Before the introduction of automated telephone exchanges, hundreds of operators worked around the clock to connect people manually. To call a friend, you would lift your phone receiver and ask an operator to connect you to the number you required. He or she connected the ringing cord with the jack on the switchboard for your friend's number. With thousands of calls made every hour, it's not surprising that lines got muddled, or crossed, and some were put through to the wrong people. Automatic exchanges multiplexed calls, where one line carried several conversations at once. In today's digital system occasional interference in the signals might mean a third person joins your call—and this is also known as a crossed line.

48 Lightbulb

ALTHOUGH THE AMERICAN INVENTOR THOMAS EDISON didn't invent the electric light, he was the first person to design an incandescent bulb that was practical to use and capable of being mass-produced.

In 1809, the English chemist Humphry Davy made the world's first electric lamp by connecting two wires to a battery and attaching a charcoal strip between the other ends of the wires. Over the ensuing decades others tried different filaments. In 1860, another Englishman, Joseph Swan, patented a bulb that used carbon filaments, but these lights often failed and very few people wanted to buy them. Edison was granted a US patent in 1880. His team worked on the project for two years and tested more than 3,000 designs before settling on carbonized bamboo as the filament. After numerous improvements he found this would glow for 1,200 hours, far longer than other materials that had been tested. However, Edison's biggest achievement was to come: Supplying the electricity to homes that would power his lightbulbs and many other electrical inventions.

Edison bulb design focused on a combination of a low electrical current, a high-resistance filament, and a secure vacuum inside the bulb. Incandescent bulbs, which glow because they are hot, were the main source of light until the 1990s, when more efficient fluorescent bulbs became the norm. These make light by electrifying very thin gases held inside.

49 Power Plant

WHEN THOMAS EDISON PATENTED HIS INCANDESCENT ELECTRIC LIGHTBULB he believed that it could replace gas lighting within a few years. But for this plan to succeed he had to find a way of generating sufficient electricity.

Edison opened the world's first power plant in Holborn, London, in 1882 to supply law offices, law courts, and the city's main telegraph office with electricity. A few weeks later Edison opened another plant at Pearl Street in Manhattan. This one boasted six "jumbo" generators, which converted mechanical power to electrical power. Steam from burning coal turned turbines, and the relative motion between a magnetic field and a conductor created the current. The generators were four times larger than any previously built. Each one

BATTLE OF THE CURRENTS

Nikola Tesla's alternating current (AC) system, with the current changing direction several times per second, was the key to sending electrical current long distances. Current could be stepped up in voltage, using transformers for long-distance transmission, then stepped down again for use in domestic appliances. Generators could be larger and fewer, and distribution costs were consequently lower. Tesla's system of generators and transformers was the decisive breakthrough needed to make electricity economical. By the 1890s, most homes and industry used AC.

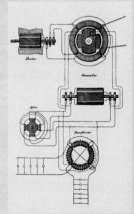

The Pearl Street power plant required a huge amount of coal to keep its generators turning and it didn't start to make a profit until 1884. In 1890, it burned down.

weighed 27 tons and produced 100 kW, enough to power more than 1,000 lights. Edison needed six generators to cover just one square mile (2.5 square kilometers) of New York. Among the first customers was *The New York Times*, whose staff reported that the quality of light in their new bulbs was "soft, mellow, grateful to the eye," compared with the harsh arc lighting they replaced. As demand grew, the failings of Edison's direct current system became apparent. This form of electricity could not be transmitted very far, so power stations were needed every 2 miles (3.2 kilometers).

Edison's rival George Westinghouse argued that alternating current (AC) was a better option. He was helped when the Serbian inventor Nikola Tesla designed a practical transformer, which Westinghouse licensed in 1888. AC power could be boosted to high voltages, which is much more efficient for sending long distances through relatively thin, lightweight wires.

Thomas Edison fought back. For several years the so-called "War of Currents" raged with claims and counterclaims. Edison insisted high-voltage AC lines were too dangerous—electrocutions were not uncommon. However, by the 1890s the two biggest AC companies were posting larger profits than Edison and he was forced to accept defeat.

Most electricity is still produced in power plants by turbines that drive generators, but there are several different kinds, some burning fossil fuels, others dependant on nuclear reactions, and some using "renewables." About 40 percent of the world's electricity comes from steam-driven turbines burning coal and oil. Natural-gas powered turbines are responsible for 20 percent, and 15 percent comes from nuclear reactors.

GEOTHERMAL POWER

The heat released from Earth could provide an almost unlimited source of geothermal energy. In some regions, rock and water near the surface of the planet is heated to up to 700 °F (370 °C). In the 20th century engineers began to tap this heat, most notably in California and Iceland (above) to produce large amounts of electricity.

50 Brooklyn Bridge

WHEN IT OPENED IN 1883, THE BROOKLYN BRIDGE was dubbed the "eighth wonder of the world," but the construction of one of New York's most familiar landmarks had been marred by a series of tragedies.

The bridge was the first to link Manhattan and Brooklyn, then two separate cities. With a main span of 1,595 feet (486 m), it was easily the longest suspension bridge in the world, and one of the first steel-wire suspension bridges. Its cables were strung between two 278-foot (85 m) towers built from limestone and granite. The clearance of 135 feet (41 m) under the roadway allowed large ships to pass underneath.

The first tragedy struck before construction had even started when the bridge's designer, John Roebling, had his foot crushed by a ferry while he was surveying. His toes were amputated but he contracted tetanus and died!

The first construction task was to build the gigantic support towers.

It took 14 months to spin the woven steel cables between the neo-Gothic towers, and a further five years to suspend the roadway from the cables.

Their foundations were to rest on bedrock deep under the waters of the East River, and Roebling's ingenious strategy was to sink pine caissons where the towers were to be. The caissons were enormous upside-down boxes pumped full of compressed air to stop them filling with water once they settled on the bed of the river. Teams of workmen inside shoveled out the sediment while the towers were built on top of the caissons. These settled lower and lower as the sediment was removed.

Some 150,000 people crossed the bridge on the day it opened, May 24, 1883. It had taken 14 years to build, at a cost of $15 million and 27 lives. The suspension design concerned some, but on May 17, 1884, P. T. Barnum led 21 elephants over the Brooklyn Bridge to prove its strength.

THE PROTOTYPE

John Roebling's reputation was built with the suspension bridge that bears his name spanning the Ohio River at Cincinnati. To lay the foundations of the two towers, timber-walled excavation pits drained by pumps let workers dig down to the riverbed. Roebling used caissons rather than excavation pits at Brooklyn, but the principle was the same. The 1,057-foot (322-m) central span of the Cincinnati bridge made it the longest of its type in the world when it opened in 1866.

51 Skyscraper

ALTHOUGH IT WOULDN'T BE CONSIDERED VERY TALL now, in 1885 the 10-story Home Insurance Building rose above the wasteland created by Chicago's Great Fire. It changed forever the way we design cities.

The modern definition of a skyscraper is an office or residential building more than 40 stories high. The Home Insurance Building rose to "just" one quarter of that, but the revolutionary design of its architect, William Le Baron Jenney, would in later years be used to construct much taller structures. Apparently, Jenney got his inspiration after seeing his wife place a heavy book on a small birdcage, which easily supported the weight. Load-bearing walls would only support buildings up to a certain height, so Jenney used steel frames, which could support much greater weights—and heights. Curtain walls could rest on, or hang from, the steel framework. His Chicago design was particularly popular because it was fireproof – a combination of a steel skeleton and stone cladding. It also boasted elevators, which had been invented by Elisha Otis a few years previously. There was keen competition between Chicago and New York, two cities where space was at a premium, for the tallest building. New York led the way for many years, until the 1,451-foot (442-m) Sears (now Willis) Tower was built in Chicago in 1973. This was only surpassed when the 1,776-foot (541.3-m) World Trade Center was completed in New York in 2014.

MUDSCRAPERS

From a distance, Shibam in Yemen looks like a modern high-rise development. It has even been called "the Manhattan of the desert." A closer looks reveals it to be very much older – and the five- to eleven-story buildings are made from mudbricks. The settlement was built this way in the 16th century as a defense against nomadic raiders. Fresh mud is regularly painted on the walls so they don't wear away.

New York's Empire State Building is one of the world's most famous skyscrapers. When completed in 1931 its 102-story, 1,454-foot (443 m) form dominated the New York skyline for 40 years before it was finally surpassed by the twin towers of the World Trade Center.

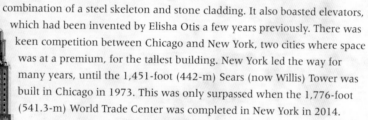

52 Automobile

FROM STEAM-POWERED VEHICLES that could move no faster than walking pace to jet-powered dragsters, automobiles have changed the way we live.

The key invention that made practical automobile transportation possible was the internal combustion engine. In 1807, Frenchman François Isaac de Rivaz invented the first engine powered by internal combustion. It was fueled by a hydrogen-oxygen gas mixture, which was ignited by a spark. Internal combustion engines were lighter, more compact, more efficient, and quicker to get started than steam engines and other external combustion engines.

The Benz Patent-Motorwagen had a rear-mounted 954 cc single-cylinder, four-stro engine. It was fueled by ligroin, a type of gasoline available in pharmacies, and ignition was by a sparking trembler coil. T steel-spoked wheels ran solid rubber tires, makir for a bumpy ride.

Powered vehicles

Throughout much of the 19th century, engineers competed to develop more efficient internal combustion engines. In 1860, the Belgian Étienne Lenoir built the first single-cylinder, two-stroke engine, which burned a mixture of coal gas and air and was ignited by "jumping sparks." He used this gas engine to power a vehicle he called the Hippomobile, which completed a test drive of 11 miles (17.7 kilometers) in three hours in 1863.

The next few years saw one new development after another in automobile technology, with most of the innovations taking place in Germany. In 1864, Otto and Eugen Langen built a four-stroke internal combustion engine, more efficient than a two-stroke, and in 1876 came the first engine that compressed the fuel mix prior to combustion to achieve far higher efficiency. In 1885, Gottlieb Daimler's Reitwagen was the world's first two-wheeler powered by an internal combustion engine; it could lay claim to being the first motorcycle.

STEAM CARRIAGE

French army captain Nicolas-Joseph Cugnot invented the world's first self-propelled mechanical vehicle, the *fardier a vapeur* or steam carriage. It was a three-wheeler, with the boiler and driving mechanism supported by the front wheel. It was built for the military but it was slow, unstable, and difficult to control. On one trial in 1771, it ran out of control and demolished a wall, the first automobile accident in history.

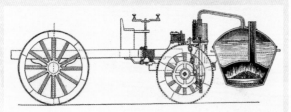

First car

Another German, Karl Benz, had long dreamed of inventing a practical "horseless carriage." After designing a gasoline-powered two-stroke piston engine in 1873, he concentrated on developing a motorized vehicle while maintaining a career as a

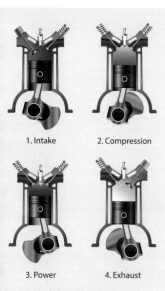

1. Intake   2. Compression

3. Power   4. Exhaust

FOUR-STROKE ENGINE

1. The piston moves down inside the cylinder, drawing an air-fuel mixture into the vacuum created. 2. As the rotating crankshaft pushes the piston up again, the air-fuel mix is compressed. 3. A trembler coil, or spark plug, sparks and ignites the fuel. The explosion pushes the piston down and the crankshaft around. 4. The piston is pushed up, forcing the exhaust gases out of the cylinder.

designer and manufacturer of stationary engines. In 1885, Benz completed the Benz Patent-Motorwagen, a three-wheeled automobile with a sedate 10-mph (16-kph) top speed.

The following year the Patent-Motorwagen was the first car to go into production with an internal combustion engine. Karl Benz was better at inventing things than he was at marketing, so to publicize his creation and start filling the order books, in August 1888, his wife Bertha took a road trip. With her two teen sons, she drove the 66 miles (106 kilometers) from Mannheim to Pforzheim in a day to visit her mother. She had to refuel at a pharmacy and the boys had to push the car up the steeper hills, but the publicity worked and Karl was soon getting orders. In 1899, his Mannheim factory was the biggest automobile producer in the world, producing 572 vehicles.

Lenoir's two-stroke engine.

Mass production

If German engineers had led the way with their inventions, it was left to an American to pioneer mass-production techniques. Henry Ford introduced production lines and rapid assembly to reduce manufacturing costs and market prices. By the mid-1920s, his Model T, of which 15 million were assembled between 1908 and 1927, was rolling off the lines at a rate of one every three minutes! The rest of the 20th century saw massive improvements in power, engine efficiency, steering, comfort, and—driven by concerns about fossil fuel sustainability—a move toward more efficient electric cars and those powered by biofuels.

Chassis at the Detroit Ford plant on the way to becoming fully formed Model Ts.

53 Turbine

IN 1894, CHARLES PARSONS' EXPERIMENTAL *TURBINIA*, a 100-foot (30-m) turbine-powered steamship, was launched in northern England. This new engine revolutionized maritime transportation and electricity generation.

Parsons called his boat the "North Sea greyhound," and in trials, the craft raced through the water at a previously unheard-of speed of 34 knots (63 kph). Three years later *Turbinia* made a dramatic appearance at the Royal Navy's review of warships for Queen Victoria's Diamond Jubilee at Spithead. In front of thousands of amazed onlookers, *Turbinia* appeared unannounced and uninvited. It raced between the two lines of naval ships and steamed up and down in front of the Prince of Wales and dignitaries from around the world. It easily outraced a Navy boat sent to intercept it.

Turbinia with the throttle open shows its speed after improvements were made in 1897.

Parsons had proved his point. His was the fastest ship in the world. The secret of its success was its three propeller-like turbines, which were spun by a stream of high-pressure steam. Each turbine was fitted to three shafts, and each shaft in turn drove three propellers, giving a total of nine propellers.

Power source

Parsons spent the next ten years perfecting the control of steam in the turbine to get the most power out of the system. His invention was used to drive a generator that produced cheap electricity, then Parsons spent the next few years developing a steam turbine that could transform marine propulsion by driving the propeller directly. In the years following *Turbinia's* dramatic appearance on the world stage, the British Royal Navy commissioned two turbine-powered destroyers from Parsons' Turbinia Works. The first turbine-powered merchant vessel, TS *King Edward*, was launched in 1901. Four years later the British confirmed that all future Royal Navy vessels would be turbine-powered. These mighty ships, or dreadnoughts, saw action in World War I.

A steam turbine used for generating electricity with a large magnet positioned above the rotating parts.

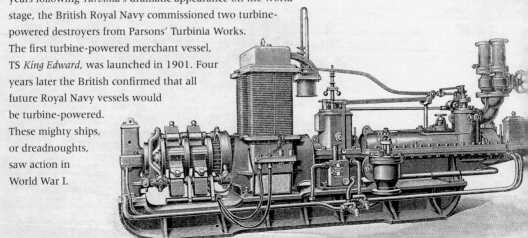

54 Radio Communication

CELL PHONES, SATELLITE COMMUNICATION, MICROWAVE OVENS, and TV broadcasts all work because of electromagnetic radiation, or radio waves. Heinrich Hertz proved the existence of these waves, but it was Guglielmo Marconi who made the first practical use of them.

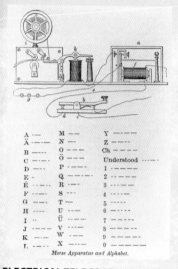

Morse Apparatus and Alphabet.

ELECTRICAL TELEGRAPH

Many people helped develop the electrical telegraph in the 1830s and 1840s, but Samuel Morse was the most famous. Electrical signals were sent over a wire from a transmitter and converted to sound by a receiver. Morse devised a language of clicks—Morse code—to convey messages. He sent a message from Washington, D.C. to Baltimore in 1844. By 1866, a telegraph wire had been laid across the Atlantic Ocean to link the United States and Europe.

The young Marconi wanted to use the "Hertzian waves," as they were known, to improve communications. With the help of his family's butler, in 1895 he invented the "wireless telegraph" in his parents' house in Italy. This was the first radio, and it could send Morse code signals about one mile (1.6 kilometers). To improve the system he needed cash; with a grant from the British Post Office, Marconi moved to London.

In 1897, Marconi set up a wireless telegraph business in England. He established a radio station on the Isle of Wight from which Queen Victoria could send messages to her son on the Royal Yacht, and he sent the first radio signal across the English Channel. His transmitter used a continuous stream of high-voltage sparks to produce radio waves, which were emitted from an antenna. The receiver's antenna detected these and they were converted to sound.

Despite Marconi's success, other engineers told him that there was a limit to how far radio signals could be sent. They argued the waves traveled in straight lines rather than following Earth's curvature, so they would just disappear into space. For his invention to be of real use, Marconi had to prove them wrong. On December 12, 1901, he received a simple message—"S" in Morse code—which had been sent more than 2,000 miles (3,220 kilometers) from Cornwall, England, to Newfoundland, Canada. The radio waves had bounced off the ionosphere (charged layers in the atmosphere) to reach their destination. This achievement brought Marconi lasting acclaim. However, it was Reginald Fessenden, a Canadian, who developed the technology to transmit voices, not just Morse code. Using an alternator-transmitter instead of Marconi's spark-gap transmitter, on Christmas Eve, 1906, he broadcast the first radio program from Brant Rock, Massachusetts, featuring a recording of Handel's "Largo" and a reading from the Bible.

Marconi poses with his early radio equipment.

55 Flyer I

PEOPLE HAD BEEN ABLE TO FLY IN HOT-AIR BALLOONS, gliders, and airships for many years, but the challenge of piloting a powered, heavier-than-air plane remained until the Wright Brothers launched their flying machine at Kitty Hawk, North Carolina, in 1903.

Flyer I was a biplane with a wooden frame, cotton wing coverings, and a simple gasoline engine. A chain drive (basically, bicycle technology) powered its twin propellers. The pilot steered with a cradle attached to his hips; this adjusted the trailing edge of the wings and twisted the rudder.

Many years after the pioneering flights, aviation experts showed that Flyer I was so unstable that no one other than Wilbur and Orville would have been able to control it.

In the 1890s, the brothers Wilbur and Orville Wright opened a store in Dayton, Ohio, where they sold and maintained the newly popular "safety bicycles." They even started building their own in 1896. The two men honed their mechanical skills on repairs to bicycles, motors, and other machines. A fascination with flight led them to experiment with gliders from 1900. Before flying a powered plane, they wanted to solve the problem of controlling unpowered ones. So they built their own wind tunnel to collect data to improve the efficiency of wings and tested 200 different wing designs. The Wrights' big technological breakthrough was inventing the three-axis control—pitch, roll, and yaw—enabling a pilot to steer and keep the plane flying.

After Wilbur's unsuccessful attempt three days earlier, on December 17, 1903, the Wrights' Flyer I made four powered flights. Orville and Wilbur took turns to lie on their stomach on the wing and pilot the craft. Orville went first and managed to keep the plane in the air for 12 seconds, during which time it traveled 120 feet (36.5 meters). The last flight of the day, with Wilbur piloting, was much more impressive: 852 feet (259.7 meters) in 59 seconds.

After the last flight of the day, a gust of wind tipped Flyer I over and it never flew again. This didn't deter the brothers. They built two more versions of their pioneering craft, and in 1905 Wilbur showed that they had truly mastered controlled flight when he flew Flyer III on a 24-mile (38.6-kilometer) circuit, staying in the air for 39 minutes.

ADER EOLE

In October 1890, Frenh engineer Clément Ader flew a plane he called Ader Eole 164 feet (50 meters) in Brie, France. An unusually light, alcohol-fired steam engine drove the propeller while Ader himself piloted the craft, which had bat-shaped wings. It only rose 8 inches (20.3 centimeters) above the ground but could be considered the first manned flight in a heavier-than-air craft. Ader later claimed to have flown it 328 feet (100 meters) but this was never verified.

56 The Panama Canal

...e final construction of the ...nama Canal included ...dders of vast ship locks ...at lifted cargo ships up ...ove sea level to make ...e short journey across ...e Isthmus of Panama ...tween the Atlantic and ...cific oceans.

IN 1913, IN WASHINGTON, D.C., PRESIDENT WOODROW WILSON remotely detonated a explosion several thousand miles away in Panama. The blast blew a hole in a dike and the Panama Canal was complete. The Atlantic and Pacific oceans were connected.

The significance of the canal was enormous. Ships no longer had to steam around Cape Horn to travel between the east and west coasts of the USA: A 48-mile (77-km) link across the isthmus of Panama cut a massive 7,800 miles (12,553 km) from a journey between San Francisco and New York. Although the canal was short, its construction was fraught. In 1882, Frenchman Ferdinand de Lesseps, who had developed the Suez Canal linking the Mediterranean to the Indian Ocean tried to cut a canal through the isthmus at sea level, a massive task. More than 40,000 workers were employed, but tropical diseases took a heavy toll: 22,000 people died before the construction company went bankrupt.

In 1903, a deal was signed between Colombia (who owned Panama at the time) and the United States to build the important trade route. US Army major George Goethals was chief engineer. More than 6,000 men drilled and blasted an 8-mile (13-km) cut through the hills that blocked the canal's path, with 3 million tons of rock and soil removed every month. Dams created two large artificial lakes in the lowlands to reduce the amount of excavation work required. At one end of the canal, three sets of locks took ships from sea level up to 40 feet (12.2 m), and three more took them back down again at the other end. The first ship made the transit from one ocean to the other in January 1914, one of around 1,000 to make the journey that year. Today, the annual figure is around 14,700 ships.

Ships that fit into the canal's locks are designated as Panamax vessels. They are hauled into the locks by electric locomotives, called mules, running along the bank on rails.

57 Tank

A WEAPON THAT COMBINES FIREPOWER, MOBILITY, AND PROTECTION is an old idea, one that became real in 1916, when 49 British "land battleships" attacked the German lines.

In February 1915, Britain's First Lord of the Admiralty, Winston Churchill, created the "Landships Committee" to investigate an engineering solution to the stalemate of World War I. "Mother," the first prototype tank, was put through its paces in early 1916, and the new petrol-fueled Mark I tank was used in battle for the first time in September of that year. The Mark I featured all-round caterpillar tracks for crossing muddy ground and trenches, the key to the tank's success in the most difficult ground conditions. French, German, and American armored vehicles were all introduced before the end of the war. Ninety years on, the tank has altered greatly, but the basic principles remain.

The Mark 1's eight-man crew was protected by thic[k] armored steel and fought with two six-pounders an[d] four smaller machine gun[s].

58 Freeway

THE "FREEWAY" WAS FIRST PROPOSED BY AMERICAN TOWN PLANNER EDWARD BASSETT in 1930. He suggested a road that offered an unhindered flow of traffic, with no traffic signals, pedestrians, or crossroads.

Controlled-access highways limit interference from other modes of transport and thus increase the speed of traffic as well as improving safety. The Bronx River Parkway, which opened in 1924, was the first road in North America with a median strip between traffic moving in opposite directions. In the 1930s, a network of autobahns was opened in Germany, and from the 1950s, freeways were built in other parts of the world as well. Other routes cross freeways on overpasses or in underpasses. One of the busiest freeways is Highway 401 in Toronto, Canada. It has 18 lanes of traffic and 420,000 vehicles use it daily, more at busy times of year. China currently has 77,000 miles (123,920 kilometers) of freeways, more than any other country, with others underway.

Elaborate intersections allow for smooth, uninterrupted merging of traffic.

59 Television

NO SINGLE PERSON CREATED THE TELEVISION. However, the biggest contribution was that of a young Scottish inventor called John Logie Baird, who on October 2, 1925, transmitted black-and-white images in his London lab.

The image was of his ventriloquist's dummy, "Stooky Bill." Keen to see what a human face would look like, Baird fetched a worker from the office downstairs, and 20-year-old Edward Taynton became the first person to be televized in a full tonal range. The images left a lot to be desired: At just five pictures per second, movements looked jerky, and with only 30 vertical bands on the screen, the definition of the face wasn't great. But it was television.

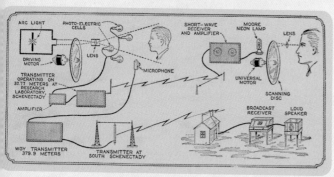

A graphic from a 1928 edition of Radio News, an American technology magazine, explains how the early TV broadcasts were put together.

Cumulative development

Baird could not have achieved this without the discoveries of several scientists who preceded him. Among these was German Paul Nipkow, whose mechanical scanning disk had small holes that picked up fragments of image and imprinted them on a light-sensitive selenium cell. In 1896, the French physicist Henri Becquerel had shown that light could be changed into electricity, and the following year German Ferdinand Braun modified a cathode-ray tube to produce images. This would later become the primary television display. Using magnetic forces in a vacuum tube to deflect beams of electrons (or cathode rays), Braun was able to produce fluorescent images on a screen.

At the same time as Baird, many engineers in the United States, Germany, the Soviet Union, and Japan were racing to create television. Baird combined a disk like Nipkow's with vacuum tube amplifiers and photoelectric cells to achieve his success. This mechanized system was used until the mid-1930s when electrical scanning techniques took over. The first color TV broadcast—the Ed Sullivan Show—was transmitted in 1951. The next significant revolution was in 1990 with the advent of digital TV. Digital TV can transmit signals very efficiently, freeing up space for more channels.

By 1935, televisions were available displaying hundreds of horizontal bands and 30 frames per second, giving a much improved image quality. All television programs were broadcast live.

60 Hydroelectricity

ELECTRICITY CAN BE GENERATED BY THE FORCE OF FLOWING WATER. The world's first hydroelectric scheme powered a single lamp, but this form of energy now supplies more than 16 percent of the world's electricity.

People have used hydropower—the ability of flowing water to do work—since ancient times. Water mills turned machinery to grind flour, for example. But it was only in the late 19th century that it became possible to turn this work into electricity. William Armstrong built the world's first hydroelectric plant (HEP) at Cragside in Britain. It powered a single arc lamp. Developments took place rapidly in North America in the 1880s and 1890s, with Thomas Edison's Vulcan Street Plant deriving its power from a turning waterwheel in the Fox River, Wisconsin.

The Hoover Dam has been generating power from the Colorado river since 1936.

THREE GORGES DAM

The Three Gorges power station, on China's Yangtze River, is the world's biggest. Its energy comes from the waters of a huge reservoir held back by the 594-foot-high (181 meters) Three Gorges Dam. When full, the reservoir is 410 miles (660 kilometers) long, and water drops 361 feet (110 meters) on its way to 32 giant turbines, which produce 22,500 MW of hydroelectricity. Work on the earth, steel, and concrete dam began in 1994, and the power station was fully operational in 2012.

Using the head

The amount of hydroelectricity that can be produced depends on the volume of flowing water and the height it falls. Engineers realized that by damming a river it would be possible to create a higher "head" of water and so increase the energy it produced. This was the inspiration behind the Hoover Dam, named for US President Herbert Hoover and overseen by his successor, Franklin D. Roosevelt. Completed in 1936, it is a curved, concrete dam 726 feet (221 meters) high and was engineered across a gorge cut by the Colorado river between Nevada and Arizona. It used enough concrete to pave a two-lane highway between San Francisco and New York. Upriver of the dam, it took six years for the gorge to fill, forming Lake Mead, which has the biggest volume of any reservoir in the United States. Water is channeled through sluices called penstocks before falling 590 feet (180 meters) at 85 mph (137 kph) to drive turbines and generate 1345 MW of electricity.

Using the same basic principles, bigger HEP projects have been completed around the world. When in 1984 the Itaipu HEP opened on the Paraná river between Brazil and Paraguay, it produced ten times the electricity of the Hoover Dam plant. And 20 years later the even bigger Three Gorges project came online (*see* box, above). There are now hydroelectric power stations in 150 countries and in some—Norway, DR Congo, Brazil, and Paraguay—this form of renewable energy provides more than 85 percent of energy needs. The five biggest power stations in the world are hydroelectric.

PUMPED STORAGE HYDROPLANT

In pumped storage plants, water flows from a higher reservoir to a lower one, passing through turbines and generating electricity on the way down. At times of low electricity demand, the water is pumped up to the higher reservoir again, using power from other sources. Overall, this system uses more energy than it produces but is a useful way of maintaining a level supply during periods of high demand.

Tidal power

Another form of hydroelectric generation is tidal power, where flowing ocean water turns submerged turbines. Tidal energy production is predictable because we can calculate tidal movements. The first tidal plant was at Rance in France, and the largest is at Sihwa Lake, South Korea. There, ten 25 MW turbines in a barrage are driven by the incoming water. Exploitation of this renewable energy source is at an early stage but the potential is huge.

61 Airship

AN AIRSHIP IS A KIND OF LIGHTER-THAN-AIR CRAFT THAT CAN TRAVEL UNDER ITS OWN POWER. The era when airships offered a relatively fast means of intercontinental transportation was abruptly cut short in May 1937 with the *Hindenburg* disaster.

Flight had been possible since the Montgolfier brothers' piloted hot-air balloon rose over Paris in November 1783. But basic hot-air balloons have a major problem: They float wherever the wind takes them and, apart from their altitude, they can't be controlled. The quest for ways to propel and steer craft once airborne proved problematic. The era of airships could be said to have begun in 1852 when the French inventor Henri Giffard flew his "dirigible," filled with lighter-than-air hydrogen, 17 miles (27.3 kilometers) from Paris to Élancourt. The airship was driven by a steam-powered engine turning a propeller. Twenty years later, German engineer Paul Haenlein advanced airship design with his semirigid framed craft, the envelope of which was made airtight by means of an internal rubber coating. The airship was powered by an internal combustion engine running on coal gas.

USS Macon *flies over New York City in 1933. This US navy airship was a flying aircraft carrier. It carried four Sparrowhawk fighters that were dropped in midair and then reattached by grabbing a "sky hook." Both the* Macon *and its sister airship, the* Akron, *crashed in the mid-1930s. The* Akron *disaster killed 73, and led to the termination of the US airship program.*

Rigid structure

More developments took place in the 1880s, including Frenchman Gaston Tissandier's electric-powered flight in 1883, then the French army airship *La France* made the first fully controllable flight in 1884, with Charles Renard and Arthur Krebs at the controls. For the first time an aircraft landed where it had taken off. *La France* was 170 feet (52 meters) long and powered by an 8.5-horsepower electric motor. The shape of its outer envelope was maintained by the pressure of the hydrogen gas within and it had a stiff keel running along its length. This

HINDENBURG DISASTER

The massive hydrogen-filled passenger airship *Hindenburg* caught fire while attempting to dock on its mooring mast in New Jersey in May 1937. Of 97 people on board, 13 passengers and 22 crew members were killed, and the airship was completely destroyed. The whole tragic episode was broadcast on live radio and marked the end of the passenger airship era. The cause of the tragedy was never determined.

HYBRID AIRCRAFT

The Airlander 10 is the world's biggest aircraft with a volume of 1.3 million cubic feet (36,812 cubic meters). It is filled with lighter-than-air helium and it is this that provides 60 percent of the lift when it's taking off. The other 40 percent comes from four 350-horsepower diesel engines, which can propel it forward at 92 mph (148 kph). It can stay aloft for five days and lift 10 tons.

kind of airship was called a semirigid. An Austrian engineer called David Schwartz built the first rigid airship, with an outer envelope supported by an internal framework, making it stronger but also heavier. It crashed on its maiden flight in November 1897, but Schwartz's idea was used by the German Count von Zeppelin, who made the most successful airships of all time.

Henri Giffard's "flying cigar" dirigible from 1852.

Aerial cruise

The "zeppelins" had an outer envelope supported by a framework of triangular lattice girders covered with fabric. The engines, crew, and—later—passengers were housed in a gondola suspended under the hull. After World War I these hydrogen-filled craft were built to be cruise ships of the air. One, the *Graf Zeppelin*, made 590 flights between 1928–37, traveling more than 1 million miles (1.6 million kilometers). In 1929, it made the first aerial circumnavigation of Earth, a mission accomplished in 21 days. Then in 1932 the *Graf Zeppelin* began five years of scheduled services carrying passengers, mail, and freight between Germany and Brazil, a journey of 68 hours.

Another Zeppelin, the *Hindenburg*, began a scheduled service between Europe and North America in 1936. After a successful first season, disaster struck in 1937 when it crashed and burned, highlighting the dangers of flammable hydrogen. This effectively ended the era of rigid airships. From the 1950s, much faster and much larger winged aircraft took their place as the means to carry passengers and freight by air. Nowadays, most airships are helium-filled nonrigid "blimps," used for filming and reconnaissance.

Airships required huge hangers to house them for maintenance.

62 Radar

RADAR SOUNDS SIMPLE; RADIO WAVES ARE FIRED FROM A TRANSMITTER at the speed of light, bounce back off an object (mostly metal ones), and are picked up by a receiver to give a location.

MICROWAVE OVEN

American engineer Percy Spencer got a shock when shortwave (or microwave) radiation from a radar he was working melted a candy bar in his pocket! He went on to invent the first microwave oven the following year, 1946. A specific wavelength of microwave can twist certain molecules in food to produce heat, which then cooks the food. The first microwave oven small enough for a kitchen was sold in 1967.

However, it took decades of experimentation to get radar that was accurate enough to be useful. In 1903, the German inventor Christian Hülsmeyer invented a Telemobiloscope for detecting ships in foggy conditions and thereby preventing collisions. Although the word *radar*, which stands for "Radio Detection and Ranging," was not used until 1940, Hülsmeyer's invention was a kind of primitive radar. It had one big problem, though: It could only measure the direction of an object, not its distance. Several technological breakthroughs were needed before radar could do this. The first of these was the invention of the oscilloscope, which could calculate range from the intensity of the reflected waves. So, specialized receivers that picked up the direction of the radio "echo" could then be used to pinpont the position and distance of objects. In the years before World War II, many nations were working in secret to fine-tune radar. In 1940, the British scientists John Randall and Harry Boot developed the cavity magnetron, which produced shorter wavelength, or microwave, radiation. This had two big advantages over previous systems: It could detect smaller objects and it operated with smaller antennas, which could be fitted in the nose of a plane.

Radar proved a decisive capability in World War II, able to see approaching enemy forces when they were still beyond the horizon. This radar array was used to guide antiaircraft fire.

63 Rocket Engine

THE ROCKET WAS INVENTED SOON AFTER THE INVENTION OF GUNPOWDER more than 1,000 years ago. Since then many people imagined how rocket power could be used to reach space, but it took a war to make it happen.

GODDARD

Having worked with solid-fuel rockets in World War I, American physicist Robert Goddard turned his attention to liquid-fuel propellants since they are denser than gases and hence require smaller storage tanks. Using a liquid oxygen and gasoline propellant, he fired his rocket "Neil" a distance of 184 feet (56 meters) in a Massachusetts field on March 16, 1926. It wasn't a long flight, but it worked—the first liquid-fuel rocket launch ever.

The first long-range rocket that could be controlled and guided in flight was the V-2, invented by Wernher von Braun. Developed as a weapon by Nazi Germany in World War II to use as a supersonic bomb intended to reach as far as North America, the secret of its success was the power of its liquid-fuel engines. V-2 rocket engines used an ethanol and water mixture for fuel and liquid oxygen as the oxidizer. This meant that the V-2 could work anywhere, even in the vacuum of space, because it did not need air to burn its fuel.

These liquids were injected into a combustion chamber and ignited at a temperature of 4,870° F (2,688° C). The hot gases that resulted blasted through a nozzle in the rocket's base, and then Newton's Third Law of Motion—every action has an equal and opposite reaction—took over: The rocket's exhaust was forced backward at high speed. The reaction to this action was the rocket being forced forward at an equally high speed.

Reaching space

The first successful V-2 test flight was in October 1942, when one was sent to an altitude of 52 miles (83.7 kilometers). The following year these rockets—which had a range of 200 miles (322 kilometers) and were armed with explosive warheads—were fired at targets in the UK, killing many people. In 1944, a V-2 was the first artificial object to pass through the boundary of space, 62 miles (100 kilometers) high. Later launches sent V-2s up to 128 miles (206 kilometers).

Today, rockets are mostly used for military purposes or atmospheric and weather research, but they are also essential for space travel. Wernher von Braun moved to the USA after the war to develop the three-stage Saturn V rocket that took men to the Moon in 1969. This 360-foot (110-meter) tall rocket is still the most powerful (and noisiest) flying machine of all time.

V-2 rocket launches from Nazi base on the Baltic ast in 1944.

64 Helicopter

IN APRIL 1944, A SIKORSKY R-4 HELICOPTER FLOWN BY THE British Royal Air Force was sent to rescue the survivors of a plane crash in a remote part of Burma. The mission was successful and the benefits of helicopters became obvious to everyone.

In the 19th century, many model helicopters were able to fly, but any machine large enough to carry a human proved a failure. In the 1920s, autogyros took to the air. These aircraft had an unpowered rotor in place of a fixed wing. The propeller engine pushed the plane along, and the rotor was spun by the slipstream, creating lift. The first true helicopter, 1936's Focke-Wulf Fw 61, looked like an autogyro, except its engine spun two rotors as well as the propeller. Three years later, the Russian-American inventor Igor Sikorsky's VS-300 was flown for the first time in the United States. It had a three-blade rotor with a diameter of 28 feet (8.5 m) and a blade speed of up to 300 mph (483 kph). It also had a single, vertical tail rotor. It was this design that became the basis for future helicopter manufacture.

Igor Sikorsky poses in a hovering R-4 helicopter.

Within two years, Sikorsky's two-seat R-4 became the first mass-produced helicopter. It passed some grueling tests flying 761 miles (1,225 km) from Connecticut to Ohio nonstop, reached a speed of 90 mph (145 kph), and attained an altitude of 12,000 feet (3,658 m). More design improvements followed with the S-51. Famously, in 1946 Sikorsky organized a point-to-point race from his heliport in Connecticut. His S-51 helicopter raced a plane, a fast car, and a train over 50 miles (80.5 km), and the S-51 won!

Modern helicopters can land and take off vertically, hover, fly backward as well as forward, and operate in congested airspace. The wing-shaped rotors create lift, and when tilted forward, thrust as well. Steering is achieved by altering the tilt of individual rotors, so they collectively push the aircraft in a new direction. This versatility makes helicopters ideal for reaching areas inaccessible to other forms of transportation. Today's Chinook CH-47F can fly at 200 mph (322 kph), and the Russian Mi-26 can carry a load of more than 20 tons or 100 passengers.

65 Jet Power

TWO GREAT PIONEERS OF ENGINEERING, FRANK WHITTLE in the UK and Hans von Ohain in Germany developed jet engines capable of powering aircraft in the 1930s. Both men's engines entered service in fighters in July 1944.

The Messerschmitt Me 262 was the first mass-produced jet aircraft when it was deployed in the final months of World War II.

Frank Whittle, a British aviation engineer, patented his idea for a turbojet engine in 1930, despite other aviation experts telling him it was not practical. A few years later Whittle helped set up a business called Power Jets Ltd. to develop his engine. Although the British Air Ministry sounded enthusiastic, it offered very little financial backing and the engine developed only very slowly.

Whittle did not know it, but in Germany Hans von Ohain was experimenting with a similar kind of engine—and he was getting more government support. Both men's first jet engines, Whittle's Power Jets WU and von Ohain's Heinkel HeS 1, ran successfully in April 1937. However, Ohain's engine was the first to actually fly a plane, the Heinkel He 178 in 1939. Two years later, Whittle's engine flew the Gloster E28/39. The development race continued neck-and-neck even then: Operational jet fighters from Germany and the UK entered service virtually simultaneously in July 1944.

Flames and compressed air

Jet engines were developed for passenger and freight use after World War II. Modern jet engines have grown in power but still work on the same principles. A fan sucks air into the front of the engine, and a compressor raises the pressure of the air. The compressed air is then sprayed with fuel, and a spark ignites the mixture. The burning high-pressure gas expands enormously and blasts out of the nozzle at the back of the engine. As the jets of gas shoot backward, the engine and aircraft are thrust forward.

JET SLED

Russian Grand Duke Cyril Vladimirovich paid the Romanian inventor Henri Coanda to build a two-seat motorized sled in 1910. Coanda powered the sled with a "turbo-propulseur" engine, a piston engine driving a multibladed blower, which exhausted into a duct to thrust the sled forward. Coanda later claimed this was the forerunner of the jet engine, but there is no evidence that there was combustion in the airstream—and no evidence that the sled ever ran, although it was claimed to go 100 mph (161 kph)! That same year the inventor built the Coanda-1910, an aircraft powered by his "jet." The plane never took off since it caught fire on the runway. Nevertheless, Coanda went on to build many successful designs of propeller-powered aircraft.

66 Digital Computer

A DIGITAL COMPUTER IS A DEVICE THAT USES A CODE of 1s and 0s to control its inner circuitry so that it can manipulate data. The first digital computers were built to perform difficult math, but, as we all know now, they can do a whole lot more.

The numbers 1 and 0 put the digits into digital computing. A digital computer's hardware is an intricate network of switches. The computer's program (or software) controls which of these switches are on and off, and these instructions ensure the computer turns inputs into desired outputs. This might be as simple as inputting numbers and outputting their sums, or as complex as inputting a stored code that is transformed into a TV program or video game.

The computers of the early 20th century were electromechanical: Mechanical switches were physically opened and closed using electrical power. In 1938, the United States Navy built a computer small enough to fit on a submarine for calculating how to fire a torpedo so that it hit a moving target. That device was hardwired, meaning it only did the

TURING

The British mathematician Alan Turing outlined his theory of a "Universal Machine," a device that could solve all problems, in 1936. It revolutionized computer theory. Turing went on to play a key role cracking German military codes in World War II, probably reducing the length of the war in Europe. After the war he worked in Manchester developing software for the world's first stored-program computer, something which he had long said was possible. In 1952, he was convicted of homosexual activity, a crime in the UK at the time. The conviction lost him his work clearance, and in 1954 Turing died by eating a poisoned apple, generally accepted to be a suicide.

ENIAC was 50 feet by 30 feet (15.2 m by 9 m) and reprogramming it was very difficult. A new program would be designed on paper and then it took several days to rewire ENIAC's switches, or relays. So was the computer programmable? Yes, but not easily!

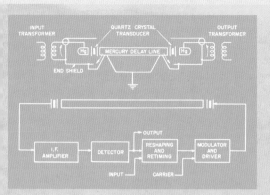

MERCURY MEMORY

In early computers, data was "memorized" in an ingenious system called a mercury delay line, which had been invented by J. Presper Eckert in the 1940s. Electrical pulses of data were converted into sound waves and sent through long tubes of mercury (which slowed down the signal). The ripples of sound were reconverted to electrical pulses at the other end (using a quartz converter) and then sent back again the other way along the mercury. Data could be bounced backward and forward inside the mercury for as long as was needed until it was retrieved by the computer. One delay line could hold 576 bits, or the code for 32 letters.

one job it was built for. A next stage, in 1941 the German Konrad Zuse built the Z3, an electromechanical computer that could be programmed to do any mathematical task.

Going electronic

By this time engineers were already experimenting with using thermionic valves as switches. These were electronic, meaning the switch had no mechanical parts; the valve could open or close a circuit depending on the way it was electrified. Electronic devices were faster than electromechanical ones. In 1946, the first digital, programmable, and electronic computer was unveiled: ENIAC (Electronic Numerical Integrator And Computer) or the "Giant Brain" as it was dubbed by the American press. Once programmed, ENIAC could perform up to 385 multiplications, 40 divisions, or three square

root operations per second. ENIAC had conditional branching: It could alter the order of execution of instructions based on the value of data. For example, "If Y is greater than 10, then go to line 26."

After ENIAC, its designers J. Presper Eckert and John Mauchly worked on a more powerful computer called EDVAC, which used the binary system (1s and 0s) rather than the decimal system (0 to 9) for the first time. One of the first commercial computers, UNIVAC 1, followed in 1951. Just 46 of these were built and each cost $1 million. The second generation of digital computers came in the late 1950s when transistors, electronics made from silicon, replaced vacuum tubes. Then integrated circuits, where many components were connected on a silicon wafer or "microchip," ushered in a third generation in the late 1960s and early 1970s. As integrated circuitry became more sophisticated, so computers became smaller yet more powerful. Today's handheld computers are billions of times faster than ENIAC.

Thermionic valves, or vacuum tubes, were a development of light bulb technology. They burned out easily, rendering the computer useless.

67 Nuclear Energy

NUCLEAR ENERGY IS RELEASED WHEN ATOMS OF THE URANIUM-235 isotope are split, creating a chain reaction. In power plants around the world, these reactions are used to produce heat, convert water into steam, and drive turbines to produce electricity.

TRINITY NUCLEAR TEST

"Trinity" was the code name of the first nuclear explosion, which took place on July 16, 1945, near Socorro, in New Mexico. "The Gadget," as the bomb was known, had a core of 13.6 pounds (6.2 kilograms) of plutonium as well as conventional explosives. It exploded with the energy equivalent of 20,000 tons of TNT and generated so much heat that the desert sand melted and turned into light green glass. Within a month, two similar bombs had been dropped on Japan, killing at least 150,000 people, forcing the Japanese to surrender, and ending World War II. Mercifully, no nuclear weapon has been used in anger since.

Our understanding of nuclear energy is relatively recent. Henri Becquerel first described how uranium emitted mysterious rays in 1896, and a few years later the Polish-French scientist Marie Curie described this phenomenon as "radioactivity." The 1930s was a decade of great atomic discovery. In 1935, the Canadian-American Arthur Dempster discovered the rare uranium-235 isotope, the only isotope existing in nature that is fissile to any extent, meaning it splits into two smaller atoms when it is hit by neutrons. Splitting the atom releases more neutrons which in turn split more atoms, so uranium-235 can produce a rapidly expanding chain reaction. This reaction creates a huge amount of energy and is the basis for the development of the atomic bomb and nuclear power.

Taming fission

Just before Christmas 1938, the German chemists Otto Hahn and Fritz Strassmann made a huge breakthrough when they succeeded in splitting uranium atoms with neutrons for the first time. On the other side of the Atlantic, at the University of Chicago in December 1942, Italian Enrico Fermi demonstrated the first nuclear chain reaction. His Chicago Pile-1 (CP-1) reactor was fueled by 5.4 tons of uranium metal and 45 tons of uranium oxide, surrounded by graphite blocks. These blocks were there to slow the neutrons to the right energy level to split the uranium. CP-1 didn't produce much energy—about enough to illuminate a lightbulb—but it was enough to prove that the reaction could be controlled.

The first industrial-scale nuclear power plant was at Calder Hall in northern England. Four nuclear reactors generated electric here from 1956 until 2003 making them the longest working nuclear reactors the world.

Atomic weapons

World War II was raging in the early 1940s, and the US government's Manhattan Project aimed to apply the power of a nuclear reaction to make an atomic bomb. There were two possible approaches: Using enriched uranium, where some of the nonfissile uranium-238 is removed to make the chain reaction faster; or to manufacture a hitherto unknown radioactive element, plutonium, from uranium so the chain reaction would be even more powerful. In 1945, the awesome power of this latter reaction was demonstrated in a fashion that couldn't have been imagined a few years before. A test detonation of a bomb with a plutonium core took place in the New Mexico desert in July. The following month a uranium bomb was dropped on Hiroshima and a plutonium one on Nagasaki. US President Harry Truman calculated that the devastation of the bombings would make it clear to Japan's leaders that they could not win the war, and save the Allies from having to risk an invasion of Japan.

...icago Pile 1 was a top ...ret facility, part of the ...anhattan Project that ...veloped the atomic bomb, ...d so no photographs of ...xist, only a handful of ...awings.

The high-speed particles emerging from the nuclear fuel create an eerie blue glow as they hit the water that cools the reactor core.

Power generation

With the war over, scientists' attentions began to turn to how nuclear reactions could be used to generate energy. In December 1951, the EBR-1 research reactor at Arco, Idaho, did just that, initially generating enough power to illuminate four 200-watt lightbulbs. Its reactor's main role was to confirm Fermi's belief that nuclear fission could actually generate more nuclear fuel. It did, and the reactor, fueled by a plutonium core, became known as a breeder reactor. It didn't produce much energy, though: Just enough to power the building it was housed in. The accolade for the first commercial nuclear power plant went to Obninsk in the Soviet Union, which opened in 1954. The nuclear fuel was encased in graphite, as in CP-1, and metal control rods were pushed into the core to soak up some of the neutrons. Lifting the control rods out made the reaction go faster, and produce more heat.

Today, about one-tenth of the world's electricity is produced by nuclear reactors. The United States and France rely most on nuclear energy for their power needs, and the biggest nuclear plant of all is Japan's Kashiwazaki-Kariwa, which produces 8,000 MW.

68 Transistor

TRANSISTORS ARE THE BUILDING BLOCKS OF INTEGRATED CIRCUITS. These are crucial components of pretty much every kind of modern electronic equipment from cell phones to heart pacemakers, computers, aircraft, and televisions.

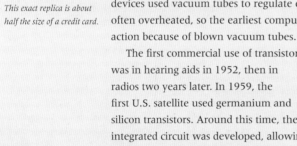

A transistor is a device that regulates the flow of current. It acts like a switch or "gate" for the movement of electrons. Modern transistors consist of three layers of semiconductor material. Semiconductors are solids such as silicon or germanium that have a high resistance to electric current—but are not so resistant that they act as insulators. A chemical process called "doping" either adds electrons to the material (creating an N-type semiconductor) or takes them away, creating P-type material (N stands for negative, P for positive). The transistor can either have an N-type layer sandwiched between two P-type layers (PNP), or vice versa (NPN). A small change in the current or voltage in the inner semiconductor layer produces a large change in the current passing through the entire component. This can switch the current on or off many times every second, making a transistor a very precise control mechanism and ideal for use in computers.

Bell Labs development

William Shockley, John Bardeen, and Walter Brattain, working at the Bell Laboratories in New Jersey, invented the transistor in 1947. The three scientists were later awarded the Nobel Prize for Physics for their achievement. Previously, computers and other complex electronic devices used vacuum tubes to regulate electronic signals, but these were bulky and often overheated, so the earliest computers were huge—and were often put out of action because of blown vacuum tubes.

The first transistor used the element germanium as the semiconductor material. This exact replica is about half the size of a credit card.

The first commercial use of transistors was in hearing aids in 1952, then in radios two years later. In 1959, the first U.S. satellite used germanium and silicon transistors. Around this time, the integrated circuit was developed, allowing a set of electronic circuits to be fixed to one silicon chip. Since then transistors have been made ever smaller. In 2008, a team of Korean engineers produced a transistor just three thousand-millionths of a meter (3 nm) across!

A microchip is a single sliver of silicon with transistors etched on to the surface in an integrated circuit. A modern microchip contains millions, if not billions, of transistors.

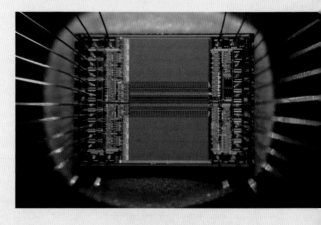

69 Laser

A LASER IS A MACHINE THAT ENERGIZES BILLIONS OF ATOMS to pump out light all at once and produce an extremely concentrated beam. Lasers perform a wide range of operations in industry, medicine, and entertainment.

In 1918, Max Planck made the discovery that was to make lasers possible—he found that light and other radiation was made up of packets of energy, which were called photons. Many years later a Columbia University scientist, Charles Townes, had the idea of a "maser," a machine that amplified invisible microwaves, and in 1957 he made a sketch of an "optical maser." A graduate student of his, Gordon Gould, applied for a patent for a Light Amplification by Stimulated Emission of Radiation device (LASER).

In 1960, Theodore H. Maiman built the first laser, based on the theoretical work of Townes, Gould, and others. Maiman used a large crystal of ruby with ends coated in mirrors of silver. Light was shined on the ruby, "injecting" it with photons. These excited the electrons in the ruby crystal. The excited electrons jumped to a high energy level, and then fell back giving out more photons. The photons bounced backward and forward between the mirrors, steadily accumulating. When the photons were released from the crystal, they formed a powerful beam of red light. The basic principle of lasers has remained the same, and the light produced can be any wavelength, or color, of light.

Lasers produce coherent light, made of waves that are all vibrating in precisely the same way. That allows the beams to be reflected very accurately. High-powered beams cut through plastics, fabrics, and even metals. They are even used as cutting and cauterizing tools in surgery. Lasers also read the data on CDs and DVDs, scan bar codes at supermarket checkouts, and provide entertainment at light shows.

Lasers can be produced from crystals, gases, and chemical reactions. They are more than pretty lights, with lasers used in medicine, beauty treatments, scanners, thermometers, extreme refrigeration systems, and guidance systems.

70 Satellite

AS PART OF HIS EXPLANATION OF THE LAW OF GRAVITATION, English physicist Isaac Newton described a cannonball that could be fired into space and circle the Earth over and over—in other words, it would be a satellite.

PROJECT ECHO

In 1962, the Telstar 1 satellite was used to transmit the first live television images between Europe and North America. However, the first communications satellite of all was a vast, shiny balloon filled with air called Echo 1. It was inflated from a gas tank placed in orbit 1,000 miles (1,600 km) up in 1960. The metallic balloon became a mirror in space, bouncing radio signals from coast to coast.

In 1957, 290 years after Newton's breakthrough, a faint beeping radio signal could be detected from the sky. This was the voice of Sputnik 1, a 185-pound (84 kg) Soviet spacecraft that was the first artificial satellite on October 4, 1957. Sputnik was a Soviet success in the Space Race, a battle between the Cold War superpowers, which was carried out in the name of furthering science although it was inspired by political rivalry. While the U.S. space agency struggled to copy the Soviet success, Sputnik 2 was launched in November, 1957, this time carrying a small dog, Laika. This showed that it was possible for animals to reach orbit. The stage was set for a human in space.

There was much more to space engineering. In 1945, English writer of both science and science fiction, Arthur C. Clarke, described a future where satellites would carry communications around the world at the speed of light. That would require putting satellites in geostationary orbits, where the spacecraft orbits at the same speed as the Earth turns, staying locked to the same position above the surface, at a height of 22,237 miles (35,787 km). This would be ideal for sending and receiving radio signals.

Unlike airplanes, artificial satellites do not have to be aerodynamic and can be virtually any shape. Sputnik 1 was an aluminum sphere with fo antenna for sending out its world-changing radio signal.

In 1946, U.S. physicist Lyman Spitzer Jr. promoted the idea of placing telescopes in low Earth orbit, some 400 miles (644 km) up. This led to the Hubble Space Telescope in 1990. Weather and spy satellites generally orbit over the poles so they can pass over almost all of the surface of Earth at regular intervals, while navigation satellites orbit in a medium Earth orbit at around 12,500 miles (20,117 km) meaning they orbit the Earth every 12 hours.

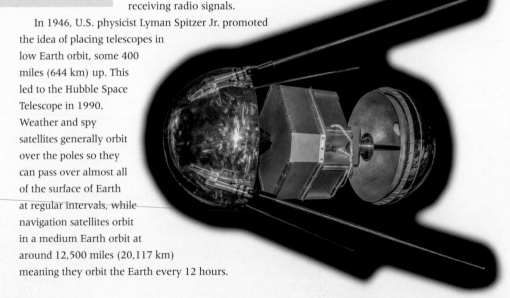

71 Fertilizers

IN 1789, THOMAS MALTHUS PREDICTED THAT THE RISING HUMAN POPULATION would eventually outstrip the Earth's capacity to grow food. Unless a new source of nutrients could be found, the human race was doomed.

The ensuing years did indeed see devastating famines and much of Earth's growing population suffered from malnutrition. However, by the 1960s, a revolution was underway—the Green Revolution—which saw a combination of technologies tackle the food shortage in developing countries. As well as new crop breeds and better irrigation techniques, the most crucial weapons in the revolution were chemicals produced on a huge scale, and largely thanks to German chemical engineer Fritz Haber.

...rtilizers, spread as ...wder or liquid, make it ...ssible to grow crops in ...il year after year without ...eeding to leave the soil to ...cover its nutrients.

Life support

All plants, including the crops we eat, need a source of nitrogen to grow since nitrogen is used to make the plants' proteins. Nitrogen is the most abundant gas in the air, but plants cannot take their nitrogen from the air, instead absorbing it in the form of nitrogen-rich compounds in the soil. When these compounds run out, the soil becomes infertile, and so farmers replace them with fertilizers. For centuries these were natural substances such as dung. In 1911, Fritz Haber developed a process for making chemical fertilizers out of thin air—or at least out of the nitrogen in it.

The Haber Process produces ammonia, a compound of nitrogen and hydrogen, which can be converted into fertilizer as well as a host of other useful chemicals, including explosives. It is estimated that between a third and a half of the world's people avoid starvation thanks to nitrogen-rich fertilizers.

HABER PROCESS

The Haber Process mixes nitrogen and hydrogen gas, heats them to around 842 °F (450 °C), then squeezes them to about 200 times the pressure of the atmosphere. In these conditions, the gases react with each other thanks to the presence of an iron catalyst. The resulting ammonia is then cooled into a liquid and drained off—and any excess gas is recycled back to the start.

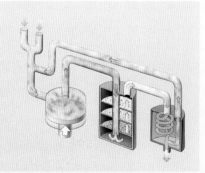

72 Oil Platform

SINCE THE FIRST WELLS WERE SUNK IN THE 19TH CENTURY, the oil industry has constantly searched for bigger reserves. Many of these were found under the seafloor but this posed serious challenges: How could drilling in the ocean be undertaken safely?

THE OIL BUSINESS

Edwin Drake drilled the first oil well near Titusville, Pennsylvania, in 1859. By 1910, major oil fields had been discovered in Sumatra, Iran, Peru, Venezuela, Mexico, and Canada. By the late 1950s, oil had become the most important fuel worldwide. The top three oil producers now are Saudi Arabia, Russia, and the United States, and 80 percent of the world's reserves are in the Middle East.

Around 1896, the first offshore drilling took place from piers built off the California coast. In the 1930s, steel barges were first used for drilling in shallow coastal waters off Texas, and a fixed platform was constructed in 14 feet (4.3 m) of water off the coast of Louisiana. Farther from shore, drilling from ships was difficult because the vessels rolled in rough seas. Technological advances allowed drilling platforms to be attached to the seafloor on concrete or steel legs, but only to depths of 1,700 feet (518 m).

Sink a little

Then something happened in the Gulf of Mexico in 1961 that revolutionized platform design. The submersible rig Blue Water Rig No.1 was being towed in a semisubmerged state to a position where it was to be eventually lowered to the ocean floor. Engineers noticed that the massive rig was very stable while still floating. That created the kernel of an idea: Oil platforms that could float but were heavy enough to stay stable in rough seas. The solution to deepwater drilling had been discovered by accident! Two years later, Ocean Driller, the first purpose-built semisubmersible oil platform was launched.

Modern semisubmersibles can operate in up to 10,000 feet (3,048 m) of water. Fixed rigs, which are anchored to the seafloor, are also stable but cannot operate in such deep water. The other alternative is a drillship. This is far less stable but can drill in water as deep as 12,000 feet (3,658 m), usually for exploration rather than production.

The sturdy columns of the oil platform are filled with water so they sink below the surface. With a large proportion of the weight located underwater, the platform is not affected by motion of the water at the surface.

73 SR-71 Blackbird

AVIATION TECHNOLOGY IS LARGELY DRIVEN BY WAR, AS ENEMIES develop
planes that can outfly their counterparts. In 1964, the fastest piloted aircraft
ever was built—not to attack the enemy but to escape it.

The development of jet propulsion in the 1940s pushed aircraft to faster than 600
mph (966 kph) and raised a new question: Was it possible for an aircraft to pass the
sound barrier? This is around 767 mph (1,234 kph), the speed at which sound waves
propagate though the air. No one knew whether an aircraft traveling faster than this
would maintain its aerodynamics and be controllable—or would it become a hurtling
object destined to break up or crash? Bullets were known to break the sound barrier,
and so a bullet-shaped aircraft, Bell X-1, was to be the first to try for supersonic speeds.
On October 14, 1947, U.S. Air Force test pilot Chuck Yeager took the controls of this
rocket-powered plane and flew into the record books. It took another seven years
or so before jet-powered aircraft could go supersonic. The chief innovation was the
afterburner, where extra fuel was sprayed into the burning hot jet exhaust, creating a
secondary thrust that pushed the aircraft through the sound barrier. In 1954, the USAF
began using supersonic F-100 Super Sabres, with a top speed of 864 mph (1,390 kph).
Since then all advanced air forces have deployed supersonic jets, the fastest generally
being interceptor aircraft built to outfly an enemy.

However, in 1964 a new kind of superfast jet was introduced: The SR-71 Blackbird.
This was an unarmed spy plane that flew at 85,000 ft (25,900 m) deep into enemy
territory. If it were detected it was designed to outrun any supersonic missiles launched
to bring it down. When needed, pilots could fly faster than Mach 3 (three times the
speed of sound). At full speed the aircraft's skin reached 500 °F (260 °C), and even the
inside of the windshield got to 250 °F (121 °C). In 1976, the SR-71 set the speed record
for a jet-powered piloted aircraft at 2,193.2 mph (3,529.6 kph).

The SR-71 Blackbird had a radar-absorbing coating and its shape helped to scatter radar. The fuselage got very hot and expanded during high-speed flights. To account for the expansion, the aircraft's joints became loose when it was on the ground—and it leaked fuel constantly.

74 The Internet

FEW PIECES OF ENGINEERING HAVE CHANGED THE WORLD AS MUCH AS THE INTERNET. Born as a security system designed for military communication, the Internet is now woven into the fabric of civilian life.

INTERNET ENGINEERS

Packet switching, the technology that created the Internet, was invented independently by Paul Buran in the United States and Donald Davies (below) in England in the 1960s. However, there was another crucial bit of engineering: TCP/IP, the rules, or protocols, that data packets use as they travel through the network. TCP/IP was invented by Americans Robert Kahn and Vinton Cerf.

In 1940, George Stibitz connected his Complex Number Calculator—an electromechanical computer—to a telephone line. At the other end was a keyboard, and Stibitz proceeded to operate his device remotely. The computer as we know it today did not exist, but the computer network had just been invented.

In 1943, Thomas Watson, the president of IBM, is famously quoted as saying: "I think there is a world market for maybe five computers." Today, there are more computerized devices connected to the Internet than there are people on Earth. However, Watson was speaking before the arrival of even the most primitive digital computers in 1946, so we can forgive him. And for several years, Mr. Watson was not far wrong. Into the 1950s, computers were vast machines, filling entire rooms and costing inordinate sums. Therefore, they were few and far between, and built as stand alone "mainframes" or electronic brains stored in the bowels of company HQs tasked with doing the mind-numbing tasks required by large corporations: Payroll, accounting, stock control, etc. No one saw any particular need to connect these devices—no one that is until the U.S. military began to roll out a computer-based air-defense system in the late 1950s.

A map of the ARPANET from 1977 shows that by then the network was connected to nodes all over the continental United States, with undersea connections to Hawaii and England.

Modems

Just like Stibitz had done, military computers were connected by dedicated telephone lines. To communicate via phone line, the computer's digital signal had to be turned into a sound—an audible one that could be carried by the phone line in the same way as a voice. This was achieved by a device called a modem—a contraction of modulator-demodulator. Early modems were designed to have a regular telephone handset plugged into

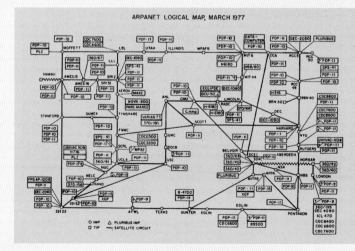

EMAIL @ SIGN

Using connected computers to send electronic messages, or email, dates back to 1961. However, the systems only worked in closed networks and needed updating for the Internet. In 1971, Ray Tomlinson devised a system that allowed messages to leave one network, or "domain," and arrive at another. He used identifying addresses with the format "name@domain." As a result @ has become a very common character. The symbol was added to the standard keyboard in the 1900s, as an abbreviation for "at." The @ sign has a longer heritage: It featured in a Greek chronicle (right) in 1345 CE!

them. The output signal went into the mouthpiece, and the inputs came out of the earpiece.

ARPANET

In the 1960s, it became apparent that the U.S. network was vulnerable to attack. One broken line could render the system useless. To solve this issue they commissioned the Advanced Research Projects Agency Network, or ARPANET. This involved connecting computers using a distributed network—something like the telephone network. Then signals sent between computers could find their own way to their destinations. If one route was blocked, the message could take another route. This made use of a communications technology called packet switching.

In packet switching the contents of a message are split up into packets. Each packet has the address of the destination computer and information about where it belongs in the original dataset. Instead of traveling together in a single unbroken stream of data in a precise order, prone to disconnections or interference, each packet travels independently and in any order. At the destination, the receiving computer reassembles the signal from all the packets, and requests missing ones to be resent.

Network of networks

The ARPANET was set up in 1969 when four U.S. universities began exchanging messages. The network grew steadily and by the late 1980s, other communication networks merged with ARPANET to create a network of networks—or the Internet. The Internet is the physical network of cables and nodes through which signals are routed according to a set of rules (known as TCP/IP). Initially, the system was used to send emails and to transfer files. However, by the 1990s, a new technology, the World Wide Web, transformed the Internet into something for everyone.

Later modems became more seamless in the way they connected, but worked the same way as early ones, and for more than 30 years were the essential component for connecting to the Internet.

75 Apollo Spacecraft

WHEN NASA'S APOLLO PROGRAM WAS LAUNCHED IN 1961, it had the aim of "landing a man on the Moon and returning him safely to Earth." Eight years later, the Apollo 11 mission achieved this aim.

On July 16, 1969, an enormous three-stage Saturn V rocket blasted off from Cape Kennedy in Florida, carrying the Apollo 11 spacecraft. The first and second stages of the rocket fell away once they were exhausted, and landed in the ocean. Then, after a brief period in Earth's orbit, the third stage of the rocket (the S-IVB) fired again. The combined command/service modules (CSM) of the spacecraft separated from the rocket, turned around, and docked with the lunar module (LM), which was inside the S-IVB. Once locked together, the CSM and LM separated from the rocket and raced through space toward the Moon.

The spacecraft had three parts. The pressurized command module was the craft's control center. It contained the astronauts' living quarters and the control panels. This was the only part that returned home. Attached to the command module was the unpressurized service module, which contained the propulsion engines, main air tanks, and propellant to push the spacecraft into and out of lunar orbit. The third part was the lunar module.

Once the spacecraft was orbiting around the Moon, astronauts Neil Armstrong and Buzz Aldrin landed the lunar module on the Moon's surface. This module had life-support systems to keep them healthy for four or five days. While they were on the Moon the CSM—commanded by the third astronaut, Michael Collins—remained in orbit around it.

After the lunar module docked with the CSM again, the craft left lunar orbit to return to Earth. The service module was jettisoned just before reentry into Earth's atmosphere. Just the command module, with all three astronauts inside, made the final phase of the journey back to Earth on its own. A heat shield prevented it burning up, and parachutes opened as it descended to its landing site in the Pacific Ocean on July 24. There were six further Apollo missions before the program was ended in 1972.

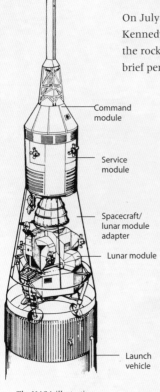

Command module

Service module

Spacecraft/ lunar module adapter

Lunar module

Launch vehicle

The NASA illustration used to explain the payload of a Saturn V rocket. The human cargo traveled in the command module, with the vehicles used to reach the Moon and land there stowed beneath.

The Apollo lunar module as presented to the public in the early 1970s, when for a short time, missions to the Moon were a routine matter.

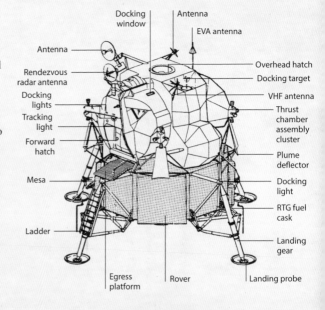

Docking window

Antenna

EVA antenna

Antenna

Rendezvous radar antenna

Docking lights

Tracking light

Forward hatch

Mesa

Ladder

Overhead hatch

Docking target

VHF antenna

Thrust chamber assembly cluster

Plume deflector

Docking light

RTG fuel cask

Landing gear

Egress platform

Rover

Landing probe

76 Jumbo Jet

AIR TRAVEL INCREASED RAPIDLY IN THE 1960s. The number of passengers rose from 106 million in 1960 to 200 million six years later. Airlines struggled to cope with the demand until Boeing unveiled its 747: The "jumbo jet."

With a "bump" at the front of the plane accommodating a double-decker section, the Boeing 747 was a distinctive plane to look at. More importantly, the capacity of the world's first widebody airliner was two and a half times that of the standard 1960s' airliner, the 707.

In service

The jumbo jet's first scheduled service, between New York and London in January 1970, was operated by Pan-Am. It was the world's first widebody airliner, capable of carrying 452 passengers in a two-class arrangement, though 550 could be squeezed in if there were no first class or business seats. It had a wingspan of 196 feet (59.7 m), and four wing-mounted jet engines gave it a cruising speed of 555 mph (893 kph). The humplike upper deck could serve as a first-class lounge or accommodate extra seating. The plane also had a cargo mode with the seats removed and a front door installed in the nose.

Sales of jumbos were slow at first, and Boeing planned to build no more than 400, but its success has gone on and on. By March 2016, more than 1,500 had been constructed and sold. The most common 747 now in passenger service is the 747-400, which has a range of 8,350 miles (13,438 km), and a passenger capacity of up to 660.

...rom the 1970s, jumbo ...ts and other widebody ...irliners have been used to ...ansport large numbers ...f passengers between the ...ggest airports, known as ...hubs." Smaller aircraft ...re then used to continue ...urneys to "spoke" ...rports.

SPRUCE GOOSE

The flying boat *Spruce Goose* (also known as the H-4 Hercules) was designed to transport heavy cargoes over the Atlantic in World War II, but the war had finished before it was completed. Built of birch wood because of wartime restrictions on the use of aluminum, it had the largest wingspan of any aircraft ever built: 321 feet (98 m), half as wide again as a jumbo jet. Although its maiden flight in 1947 was a success, it was never flown again.

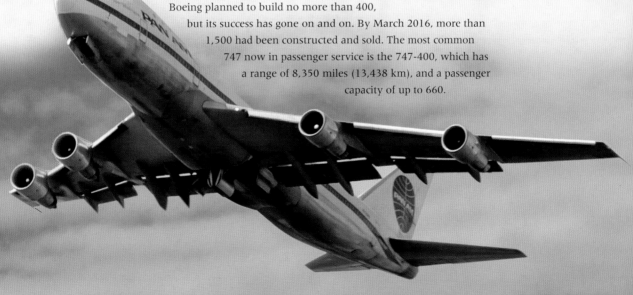

77 LCDs

IN THE 1970S, LIQUID CRYSTAL DISPLAYS, OR LCDS, became small and cheap enough to put on a watch, games machine, or pocket calculator. Today, LCDs are the most common type of screen.

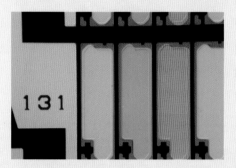

A liquid crystal is a chemical that changes its optical properties when electrified. Specifically, it blocks polarized light, a type of light beam where all the waves are vibrating in the same plane. An LCD is made up of several layers. The bottom layer is filled with LEDs, or light emitting diodes, which provide polarized white light. These pass through a layer of liquid crystals, and then on through colored filters. If light passes through all the filters the screen looks white. A colored image pattern can be displayed by electrifying the liquid crystals to the same pattern. This blocks light from reaching certain filters, so points of light of particular colors (or no color at all) appear on the display.

An LCD is made up of pixels—the bigger the screen, the more pixels, and packing in more pixels still makes any images shown sharper and more defined. Each pixel contains three subpixels: One red, one blue, and one green. Light from these subpixels combine to create dots of light that make up the image (or text) on display.

78 Genetic Engineering

IN 1974, RUDOLF JAENISCH CREATED THE FIRST GENETICALLY ENGINEERED ANIMAL—a breed of mouse that always had leukemia. Today, genetic engineers can modify the genes of all kinds of organisms, and are at the forefront of agricultural technology and medical research.

In a sense humans have been genetic engineers for several thousand years. Our crop plants, pet breeds, and domestic animals are the result of artificial breeding, where humans have changed the genetic inheritance of other organisms. However, this process takes many generations. In the 1970s, genetic engineers figured out how to short-circuit the natural process and add new DNA to organisms. There are several ways to do this: A gene gun simply blasts a sample of cells with new DNA—a few cells will absorb the DNA and survive. Other techniques make use of viruses, which add their DNA to those of their host's cells. Jaenisch created a virus that carried certain leukemia genes, and then infected mice with it. The mice's offspring inherited these extra genes.

Three mice, but none are blind. The middle one is a wild type, and the other two have been engineered with genes from a jellyfish. These genes allow the mouse to make proteins that glow in the dark.

79 MRI

IN 1977, THE FIRST FULL BODY SCAN WAS PERFORMED USING AN MRI OR MAGNETIC RESONANCE IMAGING MACHINE. Unlike X-ray scans, the images produced showed the soft tissues as well as bone.

X-ray scanners have been in use since the early years of the 20th century. They shine high-energy X rays at the body. Most pass straight through with only dense objects like bones and teeth blocking their progress. In effect the skeleton casts an X-ray shadow, which can be captured on photographic paper to form an X-ray image. X-ray scans reveal broken bones, but any damage to soft tissues goes unseen.

By the 1970s, alternatives were being developed. Ultrasound scans bounced high-frequency sound waves off internal structures, while CAT scans used X rays fired from all angles to create a more detailed image of a two-dimensional slice of the body. However, in most cases the MRI provides the best view.

MRI stands for "magnetic resonance imaging," and the scanner contains a powerful superconducting electromagnet, cooled to very low temperatures. The magnetic field it creates is about 20,000 times stronger than Earth's field. Hydrogen atoms in the body align with this magnetic field—in a harmless way. Next the machine sends a powerful radio wave through the magnetized body, making the aligned atoms flip out of position. When the radio is switched off, the atoms realign and in so doing emit their own radio signals. The MRI picks up these signals and uses them to construct an image of where the atoms are located—thus showing the internal structures in some detail. An MRI can show two-dimensional cross sections, or 3-D images.

...ery major hospital has ...MRI machine. It is used ...look at the brain, blood ...pply, and fine features of ...skeleton and muscles. ...s only used once other ...ns and tests show it ...necessary, or have not ...cceeded in diagnosing ...problem.

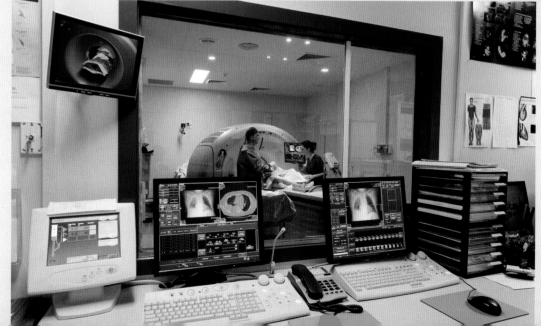

80 Bagger 288

CONTINUOUS DIGGING MACHINES CALLED BUCKET-WHEEL EXCAVATORS were built to work opencast mines. They grew ever larger until 1978 when the Bagger 288 was built. It became the heaviest land vehicle in the world.

The job the giant excavator was tasked with was truly monumental: Stripping away massive amounts of dirt and rock to get at valuable deposits of brown coal at the Hambach opencast mine in Germany. Such a big job required a big machine. The Bagger 288 is 721 feet (220 m) long and 315 feet (96 m) high. Its excavating head has a wheel 70 feet (21 m) in diameter, with 18 buckets attached. The wheel is fixed to a rotating boom, and a counterweight boom balances this to keep the Bagger from toppling over.

The Bagger (which means "excavator" in German) needs 16.56 megawatts of electricity to function. Each of the excavator's 18 buckets holds 8.6 cubic yards (6.6 cubic meters) of material, enabling it to remove 240,000 tons of coal every day. That's the equivalent of a soccer pitch dug to a depth of 100 feet (30 m), and one day's coal fills 2,400 coal wagons. One of the buckets once picked up a bulldozer by mistake!

The excavator's superstructure rests on three sets of four caterpillar tracks, each 12 feet (3.7 m) wide. The machine moves on these tracks at up to 33 feet (10 m) per minute when it has to go to a new excavation site. The Bagger 288 certainly won't win any races, but because its tracks are so big, its 13,500-ton weight is spread over a large area. It can move over earth, gravel, and even grass without leaving deep track marks. When it had completely exposed the coal source at the Hambach mine in 2001, the Bagger 288 was driven to another location 14 miles (22.5 km) away. To get there it had to travel across the Erft river, over a freeway and a railroad line, and across several roads. A team of 70 workers was needed to make sure the operation ran smoothly, and the journey took three weeks.

Bagger 288 was the first of a series of giant digging machines. It is still in use almost 50 years after it was constructed. In 2013, the mighty machine was used as part of the dystopian industrial backdrop of the movie The Hunger Games: Catching Fire.

81 *Seawise Giant*

THE DEMAND FOR OIL GREW IN THE 1970s. Refineries needed huge quantities of crude oil to produce gasoline and chemicals, and for oil-fired power stations. How could sufficient crude be transported from the oil fields?

To cope with the demand, larger tankers were needed. When the tanker *Seawise Giant* was completed at the Yokosuka shipyard in Japan, she was the biggest ship ever built, weighing 657,019 tons when fully laden and measuring 1,504 feet (458 m) long and 226 feet (69 m) wide. She was one and a half times longer and wider than the biggest U.S. aircraft carrier. Her propeller weighed 50 tons and her rudder a gargantuan 230 tons. When her 46 oil tanks were full, the bottom of *Seawise Giant*'s hull was 81 feet (25 m) below the water.

Because of a dispute between the shipbuilder and the company that had ordered her, *Seawise Giant* wasn't launched until 1981, two years later than planned. Her size created problems. She was too large to go through either the Suez or Panama canals, and she was even too big to navigate the shallow English Channel.

For several years she operated in the Gulf of Mexico, but then she went to the Persian Gulf in 1988. At that time Iran and Iraq were at war, and while *Seawise Giant* was moving a load of Iranian crude oil she was attacked by the Iraqi air force. Being so large, it was difficult to miss her! Parachute bombs landed on her deck, and fires blazed out of control. What had been the world's largest ship became the world's largest shipwreck. Incredibly, after the war was over *Seawise Giant* was towed to Singapore, rebuilt, and relaunched—this time as *Happy Giant*. For the next few years she plied the world's oceans with tanks full of crude oil. During this time *Happy Giant* was renamed *Jahre Viking*, *Knock Nevis*, *Oppama*, and, finally, *Mont*, before being scrapped in 2010.

Seawise Giant, *the largest ship ever built, being towed for repair after being attacked in 1988. Under her own steam, the tanker's powerful steam turbine engines could propel the vessel at 19 mph (30.6 kph), but it took 5 miles (8 km) for her to stop!*

82 Stealth Plane

IN 1983, A TOP SECRET MILITARY JET TOOK TO THE SKIES. Few people knew it even existed and if its design worked, no one would be able to see it coming. This was the F-117 Nighthawk, the first stealth plane.

The term stealth has a specific meaning in the military context: Technology aimed to make it impossible to identify an aircraft (or ship or submarine). The stealthy nature of the F-117 Nighthawk was largely due to the shape of the aircraft, which is why it was kept a secret by the U.S. government until 1988. (By then the U.S. air force had an even more advanced, and just as secret, stealth bomber, known as the B-2 Spirit).

A B-2 Spirit bomber has a distinctive shape, but is very hard to detect from the ground by radar. The four jets engines are concealed inside the wings, and cold air from outside is drawn in to chill the exhaust gases as they escape, thus making it hard to pick up any telltale heat signatures.

F-117 Nighthawk was designed to fly at night. In 1999, one of these stealth planes was shot down by a missile fired during the day having been targeted by sight.

While the surfaces of normal airplanes are smooth and rounded for aerodynamic reasons, the surfaces of the F-117 have many flat faces. Even the edges of the cockpit are jagged. This makes enemy radar bounce off in several directions rather than reflect back a signature pattern to the enemy. The jet engine inlets are covered by grids to disperse radar signals, and the exhaust nozzles spread the hot gases, making it harder for a heat-seeking missile to target the plane. The stealth abilities are, however, achieved at the expense of speed and maneuverability. Pilots nicknamed it "wobbly goblin."

The B-2 was a very different design, which was kept secret until 1989. Its "flying wing" design makes it faster and more aerodynamic than the F-117, but also helps to scatter radar in all directions. The F-117 was retired in 2008 but the B-2 is still in use. Stealth technology is used on warships and a new generation of military aircraft, which are made from materials that absorb radar signals or just let them pass straight through.

83 World Wide Web

TO MANY THE TERMS INTERNET AND WEB MEAN MUCH THE SAME THING.
However, the latter was engineered a full twenty years into Internet history, and made the computer network something we could all use.

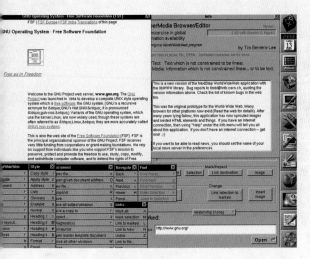

By 1989, the Internet had become a large, and growing, computer network that connected every continent (except Antarctica). It was a physical web of cables that carried data, and servers which directed the data packets hither and thither. The Internet carried emails and could be used to transfer large files from one computer to another—all very convenient. However, in today's jargon it was a "push technology." That meant the owner of the data pushed it at the other users with whom they wanted to share it. If one Internet user wanted some information from another, they might as well pick up the telephone and ask them for it to be sent—if, that is, they knew who to call.

Nexus, the world's first web browser from 1991, only worked on the now defunct NeXT computer. The back and forward buttons used in today's browsers first appeared in 1993 with the Mosaic browser, upon which modern ones are based.

In 1989, Briton Tim Berners-Lee, who led a team of computer scientists working at CERN, the nuclear physics institute near Geneva, Switzerland, devised a way of sharing information by "pulling" it through the Internet. The http, or Hypertext Transfer Protocol, made it possible for Internet users to find and view information stored on another user's computer. This created what Berners-Lee described as a World Wide Web of information. At first the information made public on web pages was only text, but web pages grew into web sites filled with all kinds of media.

Web sites are viewed using software called a browser. This loads the contents of a web site stored in some distant computer. Users navigate from one page to the next using hyperlinks: Clickable words or zones that load another web page. The hyperlink had been invented in 1945 by the American computer engineer Vannevar Bush, but it finally found its use in the Web. By the 1990s, systems were in place to search the contents of the Web, first by simply matching search terms with those on web pages. Then in 1998 Google started doing it another way…

Tim Berners-Lee is now the head of the World Wide Web Consortium (W3C) which oversees the software standards used on the web.

84 Lithium Battery

EVERY TIME WE SWITCH ON A CELL PHONE, TABLET, LAPTOP, or digital camera we are taking advantage of the amazing qualities of lithium-ion batteries. They are amazing because they are small and can be used again and again.

Ordinary "dry-cell" batteries are nonrechargeable. As the chemicals inside them are used up in chemical reactions, the output from these batteries falls and they eventually go "flat." The "dead" battery creates chemical pollution, unlike a rechargeable lithium battery (although even these wear out eventually). Although lithium batteries are initially more expensive, this is quickly compensated for by their much longer life span.

The English chemist M. Stanley Whittingham discovered the concept of intercalation—the ability to insert an ion into a layered solid such as graphite—and used this to make a rechargeable battery. The three components of a lithium battery are two electrodes and an electrolyte—a conductive medium between the electrodes through which lithium ions can move. The positive electrode is a lithium compound, and the negative electrode is usually graphite. (Pure lithium is not used because it is extremely reactive.) The magic of this kind of battery is that when it is discharging (working), positive lithium ions move from the negative electrode and enter the positive electrode. And when the battery is charging, the reverse occurs so it becomes fully charged and ready for use again.

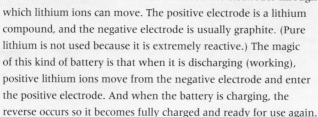

A lithium ion battery used in an electric car.

SALT MINING

Lithium is extremely reactive so does not usually exist naturally in elemental form. Lithium compounds can be obtained from hard rocks, salt lakes, and seawater. Bolivia is believed to have Earth's largest reserves, but Chile and Argentina are the leading producers. Saline water containing lithium salts is pumped from underground pools to the surface. The salts become more concentrated as the water evaporates in the Sun (below), and are then refined to produce compounds suitable for batteries.

Early attempts

Whittingham's first rechargeable battery used lithium-aluminum and titanium disulfide electrodes. Both these materials proved difficult to work with and the battery wasn't practical. Eventually, chemists settled on lithium bases for both electrodes, and the first lithium batteries were marketed in 1991. Apart from small handheld devices, lithium batteries are also now used in everything from power tools to electric cars.

85 GPS

In 1994, the final satellite in the GPS system was placed in orbit.

Although initially only available to military personnel, within a few years this space-based system would revolutionize the way we find our way around.

RADIO RANGING

Before GPS, navigators had to frequently calculate their locations. In the 1940s, a system of radio ranging was developed known as LORAN. It used a network of radio transmitters, which bounced precisely timed signals back and forth. If you knew the distance between the transmitters you could calculate your distance from them by the timings of their signals.

GPS stands for Global Positioning System. Devised as a military navigation and targeting technology by the United States, it was always intended to be put to civilian use once up and running. Today, it directs our car navigation systems, tracks our journeys and locations minute to minute, and is used to survey territory. The 1994 system used a constellation of 24 satellites (now increased to 31), which are all in a medium-Earth orbit so they arc across the sky at twice the speed of the planet's rotation. From just about any point on Earth—unless mountains or buildings get in the way—at least three GPS satellites are overhead at any one time, and can be detected by a unique radio signal.

Each signal contains the time it was sent and the location of the satellite. The radio signals travel at the speed of light, but even at that great speed there is a lag between the time sent and the time received. These tiny lags give the precise distance to each satellite. So once received by a GPS-enabled device—a car's GPS, for example—the GPS device now knows where the satellites are and how far away they are. Those data can be used to calculate the exact position of the device (and you) to within a few feet on the surface of the globe.

Space is getting more crowded with navigation satellites. The U.S.-controlled GPS system is being added to by the Russian GLONASS, European Galileo, Chinese BeiDou, and Indian NAVIC systems.

86 Optical Disc

IN 1995, A NEW STORAGE MEDIUM WAS INTRODUCED: THE DIGITAL VERSATILE DISC, OR DVD. This advance made it possible to store an entire movie on a single disc.

The disc's shine comes from a layer of aluminum foil.

The DVD was the latest design of optical disc, so-named because the data on the disc is read using laser light. The optical disc had been invented in the 1960s. The disc's surface has a spiral containing hollows or pits. Spinning the disc makes a laser run along this spiral, reflecting off the flat surface but not off the pits, so the pattern of pits creates a corresponding pattern of flickers in the laser. A detector equates the on and off laser reflections as the 1s and 0s of computer code. Early video laserdiscs and compact discs had large pits: 800 nanometers in CDs. A DVD crams in more data by having 400-nm pits arranged in two layers. Blu-Ray discs, which have superseded DVDs, have pits only 150 nm across.

MPEGS

DVDs store movies as MPEG files. These reduce the amount of data needed by only storing the part of a movie that changes from frame to frame. Everything else in the moving image simply repeats what was there previously until new data is added. This kind of file compression also makes it possible to stream high quality video through the Internet.

87 Bridge and Tunnel

THE 20TH CENTURY WAS AN AGE OF GREAT ENGINEERING. Seldom a decade passed without a new bridge or new tunnel breaking a length record. In July, 2000 a new 10-mile (16-km) crossing for the Øresund seaway in the Baltic opened making use of both: A bridge and a tunnel.

The Øresund is one of the busiest waterways in the world. It divides Denmark from Sweden and forms a crucial link between the ports of the Baltic Sea, such as Gdansk, Helsinki, and St. Petersburg, and the wider world. The Danish capital, Copenhagen, is on the western shore of the Øresund, while the Swedish city of Malmö is on the eastern side. Despite being in separate countries, together these two cities form the largest metropolitan area in Scandinavia, so a road and rail connection made a lot of sense. However, using a bridge to span the 5.6-mile (9-km) stretch of sea would block the cargo vessels that have plowed through these waters for centuries. It would also

create the conditions for ice to block the sound, generally kept clear by its turbulent currents. The solution was not simple nor cheap, but superbly elegant: A bridge and a tunnel.

Peberholm means "the pepper isle." It was given that name because it lies just to the south of the natural island Saltholm, or "salt isle."

The bridge

The Øresund Bridge is a feat in itself as a 4.9-mile (7.8-km), four-lane highway running above a railroad laid on the deck below. Supported by piles along most of its length, the largest span is in the middle, although at 1,608 ft (490 meters) it is dwarfed by the spans of some suspension bridges, which can be more than four times as wide. The Øresund Bridge uses a cable-stayed design, where the span is held by several cables secured to the bridge's towers. (A suspension bridge, like the Brooklyn Bridge, stays up because the cables are anchored to the shore at both ends, and merely hang from the towers).

Beginning on the Swedish side, the bridge never reaches Denmark. Instead, it arrives at Peberholm, an artificial island in the middle of the seaway. Peberholm is effectively a 2.5-mile (4-km) ramp that takes the cars and trains down to the Drogden Tunnel to complete the crossing.

Cost
$5.7 billion
Construction time
1995–1999
Length
Bridge: 25,738 ft (7,845 m)
Island: 13,123 ft (4,000 m)
Tunnel: 13,287 ft (4,050 m)
Traffic
17,000 cars per day

The tunnel

The Drogden Tunnel is about the same length as the island but is comprised of the world's largest concrete tubes, which were sunk into a trench dug in the shallow seabed. The spoil from this excavation makes up the bulk of Peberholm.

The bridge and tunnel link took seven years to build and cost $5.7 billion.

88 ISS

Since 2000, the International Space Station (ISS) has been a habitable laboratory, observatory, and factory orbiting more than 200 miles (322 kilometers) above Earth. It is a joint enterprise between the space agencies of the United States, Europe, Russia, Japan, and Canada.

The first component of the ISS was put in place in 1998, and since then lots of other modules have been added. The first resident crew (Expedition 1) arrived on a Russian Soyuz craft on November 2, 2000, and since then the station has been occupied continuously, usually by a crew of six scientists. They have conducted a wide range of tests, including how the human body performs in a "weightless" environment and how machinery performs in space. The ISS has been visited by astronauts and space tourists from 17 countries. The station has been serviced by visiting spacecraft, including

The ISS with a space shuttle docked to the NASA part of the space station.

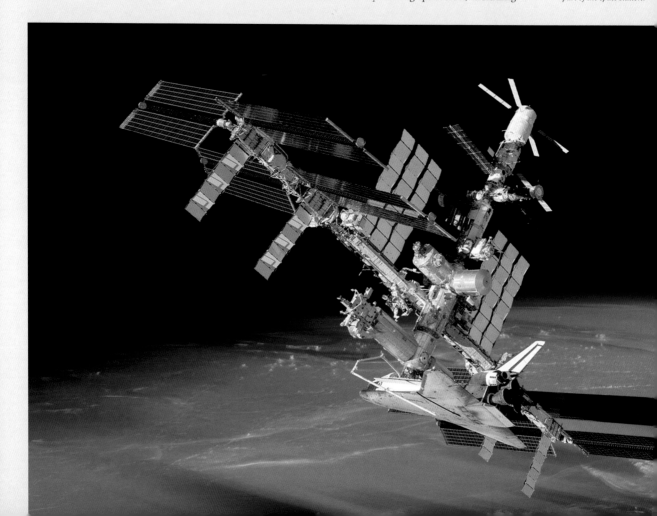

SALYUT

Salyut 1 was the world's first crewed space station. It was put into orbit by the Soviet Union in April 1971 and crewed by a team of three in June. The trio stayed on board for 23 days, keeping in regular contact with mission control. Tragically, all three died of asphyxiation on their return to Earth. There were several more Salyut missions until the program was halted in 1986. The last of these, Salyut 7, was in orbit for eight years and ten months and was visited by ten long-duration crews.

NASA's space shuttles, Russian Soyuz craft, and SpaceX's uncrewed Dragon supply vessels.

Space story

The first module of the ISS, called Zarya, was launched on a Russian Proton rocket in November 1998. Zarya provided propulsion, communications systems, and electrical power, but lacked long-term life-support systems. Two weeks later the space shuttle Endeavour brought another module, Unity, which was attached to Zarya by spacewalking astronauts. At this stage the space station did not yet have a crew. In July of the following year Zvezda docked with Zarya-Unity. This Russian module contained sleeping quarters, a toilet, kitchen, exercise equipment, and oxygen generators—enough life support to allow permanent habitation. Zvezda's engines are also used for reboost maneuvers that keep the ISS in orbit between 205 and 270 miles (330 and 435 kilometers) up. It orbits Earth 15 and a half times every day.

Later, sections of the integrated truss structure were fitted. Various devices were attached to the truss, including solar arrays to provide the main source of energy, and radiators to get rid of excess heat. More units were added to the ISS until the Columbia shuttle disaster in 2003. After an enforced break, construction started again in 2006 until there were 15 pressurized modules and an enlarged truss structure. There are plans for five more modules.

The energy needed to run the station comes from solar power, with backup from rechargeable batteries for the 35-minutes in every 92-minute orbit when the Sun is eclipsed by Earth. Food is vacuum-packed, and eaten with a knife and fork, but these are attached to a tray with magnets to keep them from floating away! Drinks are sipped from plastic bags with straws.

LIVING IN SPACE

Astronauts on the ISS sleep in sleeping bags, often in their own small cabins. In this weightless environment they must attach the bags to a wall to keep them from floating around! Shutters are pulled over windows to create a formal "nighttime", although because of its orbit speed, the ISS actually experiences 16 periods of daylight and 16 "nights" of darkness in every 24-hour period.. The standard working day is ten hours on weekdays and five hours on weekends.

89 Palm Islands

SEAFRONT LAND IS AT A PREMIUM IN DUBAI, a petroleum-rich city on the shores of the Persian Gulf. In this booming metropolis, busily reinventing itself as a tourist haven for when the oil eventually runs out, the answer is simple—just build more coastline.

Dubai is one of seven emirates—each ruled by a Arabian prince, or emir—that form the UAE, or United Arab Emirates. These territories make up a tiny slice of the Arabian Peninsula, and Dubai itself has just a 40-mile (64-kilometer) coastline. In the 1990s, a plan was hatched to change all that by adding huge artificial islands.

To maximize the length of these man-made coastlines, the engineers looked to nature. The answer was to make them in the shape of palm fronds, with multiple leaflets creating mile upon mile of private beaches and luxury resorts. Three islands were planned, which would add 360 miles (580 kilometers) to the Dubai waterfront. In 2007, Palm Jumeirah was the first island to be completed, a second is near completion, while the third has been converted to a peninsula development.

Palm Jumeirah is made from a staggering 3.3 billion cubic feet (94 million cubic meters) of sand and rock, making it the world's largest artificial ocean island. The original plan was to use desert sand (of which Dubai has a plentiful supply) but seafloor sediments proved easier to work with. The dredged sand was sprayed into mounds, using GPS tracking to create the shape. These foundations were covered in plastic mesh, and topped with rock and more sand. The sand was then vibrated with water jets to make it settle like cereal in a box.

The fronds of Palm Jumeirah are surrounded by a circular breakwater that protects the island from storms. Small gaps were added to the breakwater to allow currents to circulate around the fronds.

90 Smartphone

A SMARTPHONE IS A HANDHELD DEVICE THAT COMBINES a cell phone and a personal computer. Modern smartphones feature access to the Internet, the ability to send emails, digital cameras and videos, media players, and a GPS navigation unit.

There have been some amazing changes in the way we communicate with each other in the last three decades. Until the early 1980s, all phones were connected by wires. Then a cordless cell phone was launched by Motorola in 1983. This used radio technology to exchange voice signals with a network of antennas. Each antenna covered a zone, or cell, of territory. That meant a caller could keep moving, and his or her phone call would automatically switch to the neighboring cell. The antennas were connected to the wired telephone network, which carried the call to the nearest antenna covering the cell where the other caller was located. Early cell phones were large and heavy—and could only make calls. Today, we might call them dumb phones.

SIMON COMMUNICATOR

The first smartphone was the Simon Personal Communicator, which was invented by Frank J. Canova Jr. and launched in November 1993. This 18-ounce (510-gram) handheld cell phone and personal digital assistant (PDA) featured a touchscreen with icons you could tap or poke with a stylus. As well as being able to make phone calls it could send and receive emails and faxes. The Simon also had an address book, calendar, and calculator.

Getting smarter

In 1993, IBM launched the Simon, the first smartphone (though no one had thought up that term then). Cell-phone technology was advancing quickly. So-called second generation networks were built, where signals were handled in a digital format. Phone handsets became smaller with better batteries. The first MP3 music players were added to phones in 1999, then came camera phones (with constantly improving image quality). Internet access became a feature of 3G or third generation phones at the start of the 21st century.

When the Apple iPhone was launched in 2007 it revolutionized personal communications again. This was the first commercial cell phone with a multitouch interface: Its touchscreen recognized the presence of more than one point of contact so the user could pinch to zoom in or out, for example. From 2012, access to high-speed mobile broadband became standard. Engineers are now trying to develop foldable screens, so smartphones can be smaller and lighter. And the phones of the future may recharge batteries via radio waves rather than having to be plugged in.

A ubiquitous feature of a smartphone are soft keys. Instead of having a physical keypad and buttons, the user controls the phone by touching images on the screen. They change according to what the phone is doing—numbers for dialing, letters for typing or icons for launching apps.

91 Drone

MILITARY AIRCRAFT HAVE ALWAYS BEEN EXPENSIVE AND VALUABLE ITEMS.
However, the most valuable part of the aircraft is always the human pilot.
Drones, or unmanned aerial vehicles (UAVs), do not have pilots, and so can
be sent into action for longer and with less regard to risk.

POLICING DRONES

Civilian drones—often used as
toys—are a growing problem in
urban areas. They can stray into
the path of aircraft near airports
and be used to view private and
protected areas. In Japan, police
use their own UAVs, dragging nets
to bring down drones. However,
in the Netherlands, police deploy
trained eagles!

The advantage of drones was obvious from the early days of aviation. The
first full-sized drone was the V-1 flying bomb, or "doodlebug." This World
War II weapon, developed by the Germans, flew in a straight line for a
predetermined time and then dropped from the sky on to its target.

The next generation of drones were little more than model aircraft,
equipped with surveillance equipment. They were controlled by radio from
the ground, but this limited their range. Drones grew steadily in size until
they were the size of crewed aircraft with connections to their pilots via
satellite links. From the ground, a crew could fly the drone over any point
on Earth—and get a live bird's-eye view of the scene below. One of the
first of these drones was the Predator, which entered service in the 1990s
and could stay aloft for 24 hours at a time.

Robot plane

The introduction of the Reaper drone in 2007 marked the next step in
drone technology. This machine was faster and more powerful than the
Predator, but it could also fly autonomously, using its own sensors to
navigate along a preset course, without the need for a human pilot back at
base. This technology is now beginning to have an impact in the civilian
and leisure use of drones. In 2014, the Chinese-made Ehang UAV was
unveiled. Still in development, the E-Hang will be a passenger drone that
can carry a single human occupant on a 20-minute flight at 60 mph (97 kph). The
human on board has no flight controls and just chooses the flight path and destination.
The drone flies using eight rotors, each with a separate power supply to reduce the
dangers of engine failure.

*General Atomics MQ-9
Reaper can take off and
land under remote control
or autonomously. The UAV
can also be launched in
midair when dropped from
a piloted aircraft.*

92 E-Ink

IN THE LATE 2000S, A TECHNOLOGY CALLED E-INK AND ELECTRONIC PAPER WAS TAKING THE BOOK WORLD by storm. After many years of trial and error, electronic devices designed for reading text that were the same size and same weight as a paper book had become widely available.

The idea of an electronic device that displays the text of a book is a compelling one. It could show any page of any book at any time—an obvious advantage over printed paper. With the proliferation of smaller and smaller computing devices over the 1980s and 1990s, why did it take so long for screens to begin to replace paper? Some answers seem obvious: You don't need a power supply to read a paper book, nor do you need to connect to a computer to fill it with words.

Electronic paper

It took a while for engineering to solve these issues with better batteries and the arrival of almost ubiquitous wireless connectivity. All this was in place by the mid-2000s, but there was one final issue to resolve: The LCD displays used by computers, tablets, and phones glow—or more precisely, they transmit light into your eyes. That technology is heavy and it is taxing on the eyes. Paper is lightweight and you see the text because ambient light reflects off the page, making it much more pleasant to pore over for long periods—basically, what you have been doing since you bought this book!

An e-reader works well for reading static text. However, the e-ink screen updates very slowly compared to LCDs and so is not suitable for showing moving images or for use as an interactive device.

By 2008, an alternative technology was perfected. It was known as electronic paper. Like real paper it made use of reflected light, but was able to change the text displayed. It did this using minute black-and-white particles held in tiny oil-filled capsules. A typical e-reader screen has about 800,000 capsules. The capsules are then given a positive or negative electric charge. A negative charge repels the negative black particles, pushing them up to the surface of the screen to make a black dot. A positive change creates a white dot. So creating a pattern of charge in the screen creates a corresponding pattern of black-and-white areas. The black dots combine to show text, and the white ones create the white space.

Electronic ink is made up of tiny black-and-white particles, each about a micrometer wide.

93 Burj Khalifa

WHEN IT COMES TO THE TALLEST STRUCTURE there is a fiercely argued set of rules that has resulted in several different lists for buildings, freestanding towers, masts, etc.—but in 2010 Burj Khalifa topped them all.

The Burj Khalifa actually became the world's tallest building in 2007, long before it was finished. When it was finally opened in 2010, the tower was soaring more than half a mile (0.8 km) above the cityscape of Dubai.

Desert flowering

Taking just six years to build, the design of the mighty tower (*burj* means "tower" in Arabic) was inspired by the shape of a desert flower—the spider lily. Initially designed to reach 2,651 feet (808 m), the side view was deemed too inelegant. There was a simple solution—make it even taller! The full height is 2,722 feet (830 m) across 163 floors. It took more than 110,000 tons of concrete, 55,000 tons of reinforced steel, and a team of 12,000 workers to put it all together. (Even so it looks like the skyscraper's record is on borrowed time—in 2020, the Jeddah Tower in neighboring Saudi Arabia is destined to top out at 3,280 feet or 1 km tall!)

The Burj Khalifa has three wings, which get narrower with height. That is what makes the building strong enough to be so tall. In effect it is several skyscrapers of varying heights clustered together. As well as creating a stable structure, the petaled design also increases the area of the outside wall, so residents are never far from a window. The exterior glass would cover 20 football fields and most of the 24,348 windows have automatic washing systems. However, the upper 54 floors have to be cleaned by hand, taking four months to complete!

High rise

The mighty skyscraper is a self-contained town. It can house 35,000 people in 1,000 apartments, offices, and hotels. A record-breaking tower needs impressive elevators, and the one that goes from ground level to the 140th floor travels at 33 feet (10 meters) per second. A journey to the public observation deck on the 124th floor takes little more than a minute. As with all skyscrapers there is stair access to every story of Burj Khalifa. The total number of steps to the top is 2,909. In 2011, Alain Robert, a French free climber known as "Spiderman," took six hours to climb up the outside. Soon after, Nasr Al Niyadi and Omar Al Hegelan leapt from the 160th floor to make the world's highest BASE jump.

The spire is visible from 60 miles (97 km) away.

94 Fuel Cell

THE IDEA OF CONVERTING CHEMICAL ENERGY into electricity—with water the only waste product—is not new. But now this simple idea is being applied to generate electricity for homes, emergency power, and electric vehicles.

As long ago as 1838, the Welsh physicist William Grove built an experimental fuel cell, but it wasn't until the 1950s that the first commercial cells were manufactured. The American scientist Roger Billings designed a fuel cell that could be used to power an electric car in 1991, and by 2010, several fuel cell vehicles were in use.

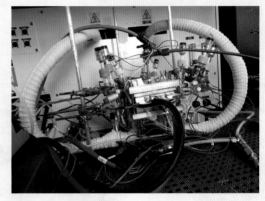

An experimental fuel cell used in research to boost efficiencies.

There are different types of fuel cell but all have a positive and negative electrode (an anode and cathode, respectively) and an electrolyte between them. Most cells are fueled by hydrogen and oxygen. Hydrogen atoms enter the anode, where they react with a catalyst (usually platinum powder). Each hydrogen atom loses its single electron, which passes through an external circuit, forming an electric current. The hydrogen atoms now have a positive charge and they move through the electrolyte to the cathode. At the cathode, the hydrogen ions react with the electrons and oxygen to form the waste product—water, which is drained from the cell. The choice of electrolyte is important: It must only allow hydrogen ions to pass through. The electrolyte may be potassium hydroxide, phosphoric acid, or a salt compound.

Unlike batteries, fuel cells require a continuous source of fuel (usually hydrogen) and oxygen to sustain the chemical reaction. However, because they have no moving parts and do not involve combustion, in ideal conditions they can achieve up to 99.9999 per cent reliability. They have another advantage, too: They don't produce pollution.

One of the most exciting uses of fuel cells is in electric cars. Several models are now in production, including the Toyota Mirai, which has a range of 312 miles (502 km), a refueling time of three to five minutes, and can go from 0 to 60 mph (97 kph) in nine seconds. Japan already has a "hydrogen highway," equipped with hydrogen fueling stations.

ASA employs fuel cells for making power in many of s spacecraft. Here a fuel ell is being removed from he payload bay of a space huttle.

95 Self-Driving Car

PEOPLE DREAMED OF CREATING DRIVERLESS CARS as long ago as the 1920s, but only since the advent of computers has this dream become reality. In 2011, Nevada became the first U.S. state to allow these cars on public roads.

Work on a computer-controlled, or autonomous, vehicle began at Carnegie Mellon University in Pittsburgh in 1984. Several big manufacturers, including Mercedes-

Sitting in the front seats of a self-driving car would be an unusual experience. With the controls stowed away, every person on board is a passenger.

Benz and General Motors, built their own prototype autonomous vehicles. Then Sebastian Thrun, the former director of Stanford University's Artificial Intelligence Lab, was hired to set up Google's self-driving car project.

Google now has a fleet of production and custom vehicles with "Chauffeur" software fitted. The vehicles have GPS and follow preprogrammed routes. Camera sensors recognize road signs, traffic signals, other vehicles, and pedestrians. Lasers calculate distances. If a red signal shows, the car's sensors detect it and stop at the appropriate place. It keeps a safe distance from other vehicles at all times. The onboard computer has a manual override so a human driver can take over at any time. By March 2016, a fleet of these autos had clocked up 1.5 million road-miles (nearly 2.4 million kilometers), many through heavy city traffic. Although there had been a small number of accidents, they were mostly the fault of other drivers.

Self-driving cars offer independence to people unqualified to drive through age or disability and they free up journey time for other tasks. They should also reduce accidents and allow better control of traffic flows. Experts predict that by 2040 up to 75 per cent of all road vehicles will be autonomous.

Google has led the development of self-driving cars. The dome on the top of this early model contains the sensors that tell the car what is happening around it.

96 Curiosity Rover

IN THE ABSENCE OF HUMAN EXPLORERS, OUR INVESTIGATIONS OF MARS have been carried out by a team of robot rovers, growing in complexity. In 2012, the latest explorer, Curiosity, touched down.

Curiosity is the fourth rover to roll around Mars. The first was Sojourner, a six-wheeled truck not much bigger than a remote-controlled toy, that bounced down inside airbags in 1997. Sojourner proved that solar-powered rovers could work on the red planet, and two larger rovers—Spirit and Opportunity—were sent to Mars in 2004. Like Sojourner they were solar powered, and engineers were not sure the rovers would survive their first dark Martian winter without losing all power. As winter approached, the rovers were parked on slopes facing the low, weak Sun. With careful power management, they remained operational. In 2009, Spirit became stuck in deep sand, and, unable to move to a winter refuge, finally lost power in 2011. However, Opportunity is working to this day, outperforming its 90-day primary mission many times over. In 2012, it was joined on Mars by Curiosity, an altogether different vehicle.

Curiosity is the size of a family car, four times the size of Spirit or Opportunity. It is powered by a radioactive electrical generator, which could provide power for two decades or more throughout the Martian year. Curiosity also arrived on Mars in a different way. Instead of bouncing down in airbags, a "skycrane" hovering under rocket power lowered the 1,982-pound (899-kg) craft to the surface. Like all Martian rovers, Curiosity is a mobile geology lab. Equipped with drills, shovels, and all kinds of miniature detectors, it is moving across Mars's Gale Crater analyzing the minerals and larger scale rock formations. This kind of information not only tells NASA scientists about the natural history of Mars, but also helps plan the first human mission to Mars.

Curiosity's stereoscopic video cameras create a 3-D image of its surroundings, which controllers back on Earth use to guide it across Mars. The robot arm is used to analyze rocks—searching for signs of water, and perhaps ancient Martian life forms.

97 Atlas Robot

ATLAS IS ONE OF THE MOST ADVANCED COMPUTER-CONTROLLED ROBOTS. This humanoid robot is designed to aid emergency services in search-and-rescue operations in environments that are too risky for humans.

The original Atlas was 6 feet (1.82 m) tall and weighed 330 pounds (150 kg). The latest one is 5 feet 9 inches (1.75 m) tall and weighs 180 pounds (82 kg).

Imagine the scene of an earthquake where buildings have collapsed, deadly chemicals are leaking from shattered pipes, and fires are raging. This kind of nightmare situation is one that even the best-equipped and trained emergency teams struggle with. Enter Atlas the robot. Atlas was financed by DARPA, an agency of the U.S. Department of Defense, and developed by Boston Dynamics and other corporations. When it was first unveiled in 2013 the range of skills it could perform amazed the assembled guests. Atlas is made of aluminum and titanium, and is electrically powered and hydraulically actuated. Its hands have advanced motor skills, and all four limbs have joints that can move independently in several directions. The robot's onboard computer controls stereo cameras to evaluate the world around it. In this it is helped by laser distance detection (lidar).

Skill showcase

To show off the latest robot technology and encourage even better design and innovation, DARPA organizes a Robotics Challenge. In 2015, Atlas performed spectacularly in this, accomplishing many of the tasks an emergency team would have in a simulated life-threatening situation. Atlas drove a utility vehicle, clambered across a pile of rubble, removed debris blocking the entrance to a building, and opened a door to enter the building. Then it climbed a ladder, used a tool to break through a panel, closed a valve to shut off a leaking pipe, and finally connected a hose to a standpipe and turned on the water.

The latest, improved version of Atlas can walk on a greater variety of surfaces, including snow and icy slopes. If knocked, it rights itself, and if it falls over, the robot can stand up again.

98 Wind Power

ATTENTION TURNED TO RENEWABLE FORMS OF ENERGY as concerns grew about the impact on the climate of burning fossil fuels. Wind power is now the second-biggest provider of electricity from renewable sources, after hydropower, and is increasing every year.

Wind turbines convert the kinetic energy of wind into electrical energy. Most turbines consist of a steel tower with two or three blades attached. When the wind blows, the blades rotate, turning a shaft that runs into a streamlined box called a nacelle. There, a gearbox accelerates the rotation of the shaft, and a generator converts the energy of this movement into electricity. Sensors monitor the wind direction and swing the blades to face into the wind, so they receive the maximum benefit from it.

A wind turbine is really just the latest incarnation of the windmill. In the first century CE, Heron of Alexandria designed a "windwheel" to blow air through organ pipes, and windmills were used in Persia to pump water from the ground from at least the 9th century. In the Middle Ages their use spread to India, China, and Europe. In 1887, James Blyth built the first windmill to generate electricity in the backyard of his Scottish vacation cottage.

Denmark was a pioneer of wind turbines, with windmills contributing to the nation's energy supply from the early 20th century. But the technology has only developed globally in the last three decades. Worldwide there are now more than a quarter of a million wind turbines. Groups of turbines, or "wind farms," are sited in windy locations, some on land and others offshore. The largest project is at Gansu, which helped China become the world's biggest generator of wind power in 2013. When complete, Gansu will produce 20 MW of electricity. The Vestas V164 is the world's biggest turbine. Operating off the coast of Denmark, its blades are 538 feet (164 m) in diameter and it stands 722 feet (220 m) high.

Offshore wind farms are more efficient because the flow of ocean wind is more constant, unperturbed by hills and other landscape features.

99 Maglev Train

WHEELS HAVE BEEN THE UBIQUITOUS FEATURE OF LAND TRANSPORTATION FOR at least 7,000 years. Before that people simply walked. However, the age of the wheel may be coming to an end.

FLOATING MAGNETS

Magnetic levitation is achieved when the repulsive force on a magnet equals the gravitational pull—so instead of falling down, the magnet floats in midair. The effect is most stable using superconductors that are chilled to low temperatures. Seen above, a magnet is floating over a superconductor chilled by liquid nitrogen. The magnet's field creates corresponding electric and magnetic fields in the superconductor, which hold the magnet in place.

The term "maglev" is short for "magnetic levitation," and although maglev vehicles are still very much a futuristic concept, the ideas behind them are 100 years old. In 1911, Dutch physicist Heike Kamerlingh Onnes used the world's most powerful refrigerator to cool materials down to incredibly low temperatures, close to 0 Kelvin or −459.67 °F (−273.15 °C). This is the lowest temperature possible, known as "absolute zero." When Onnes cooled mercury to near this level he found that its electrical resistance faded to nothing. Resistance is a property where a material stops electricity from flowing through it. Some materials have a high resistance; others, such as copper wires, have a low resistance. Super-cold mercury has no resistance at all! This was the discovery of superconductivity, where materials could be used to carry electricity without wasting any energy. Since then engineers have developed a whole range of superconducting materials, and discovered they are capable of other very strange behaviors. Although they require very cold temperatures, superconducting magnets are very powerful, powerful enough to make a whole train levitate off the ground. If this phenomenon could be used in transportation, vehicles could float along at high speed, unhindered by friction.

The Shanghai Transrapid is the only high-speed passenger maglev service in operation to date. Using a monorail system, the train has a top speed of 270 mph (435 kph) and carries passengers from downtown Shanghai to the airport 19 miles (31 km) away in about 8 minutes.

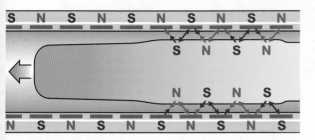

The propulsion system of a maglev train uses the same magnets that keep the train hovering. By switching the polarity of the magnets, the system creates a wave of pushing and pulling forces along the train that accelerates it to high speeds.

Magnetic track

While hovering cars is an appealing prospect, it was the railroad that was best suited to develop into a maglev system. The first tracks, trains, and a few short maglev lines were built in the 1960s. These early systems were monorails where the train cars hooked around a central rail. Powerful magnets under the track pulled similar magnets on the wrapped-under part of the train. That attraction raised the rail cars a fraction of an inch above the track. The system worked but was far from efficient and very costly to maintain. Most of the lines have been shut down.

However, the dream of maglev lines did not die. Instead, a new electrodynamic track is being developed where the train levitates, nestled inside a trackway, due to the repulsion of the magnets lining the track and those under the train. This arrangement of magnets not only makes the train levitate, it is also used to push the train along.

Linear motor

The motor that makes a maglev train move has no moving parts. Instead, it uses magnetic forces in a system known as a linear motor. As is common knowledge, the opposite poles of a magnet attract, while like poles repel. So to make the train levitate, the polarity of the track is opposite to the polarity of the train. However, to make the train move, the polarity of the magnets is switched back and forth rapidly. This creates a wave of attractive and repulsive forces that travel down the trackway, pushing and pulling the train along. Because there is no friction between the wheels and the rails of a track, once the train has started moving there is little to slow it down.

Record breaking

Most maglev research takes place in Germany and Japan. In recent years, Japanese engineers have been taking the lead, as their experimental trains have pushed up the record speed for train travel. In 2015, a maglev train become the first to exceed the 600 kph barrier—or 373 mph. While there is no friction with the track, at these speeds air resistance does become a limiting factor. The fastest maglev trains are designed with long, streamlined noses so they slice through the air smoothly. At the moment these trains only run on short test tracks and there are no plans to develop long-distance maglev systems. The cost of such an engineering project would be very high. However, if engineers can develop less expensive superconductors, perhaps levitating transportation will become something we can all use.

TGV

The Train à Grande Vitesse—French for "high-speed train"—is better known as the TGV. The TGV is the model for most high-speed rail networks across the world. One exception is the Japanese Shinkansen—or "bullet train"—which showed the potential of fast rail in the 1960s. Since the 1990s, the fastest trains have been hauled by the electric motors of the TGV. On a regular journey, the cars cruise along at nearly 200 mph (322 kph). In 2007, a test train reached 357 mph (575 kph) setting a new record for trains running on rails. Only specialist maglev trains have reached higher speeds.

100 Solar Power

SOLAR POWER IS THE CONVERSION OF SUNLIGHT into electricity. It is the ultimate form of renewable energy because there's an endless supply of sunshine and it does not create greenhouse gases.

SOLAR FURNACE

A solar furnace uses parabolic mirrors called heliostats to focus sunlight on a small area to produce extremely high temperatures. The largest is at Odeillo, in France, where 63 heliostats concentrate sunlight on a small target zone, creating temperatures up to 6,330 °F (3,500 °C). The installation is used to research nanotechnology and how materials react to thermal shock—rapid increases in temperature. Of course, solar furnaces only work on sunny days.

In 1974, only six homes in North America were heated or cooled by solar power. Oil was then a cheap form of fuel. When oil prices later increased, engineers began to look at ways to make solar power a practical form of energy. Research into solar energy got more serious when concerns were raised over the impact of burning fossil fuels on the world's climate. Experts now forecast that by 2050 solar power will be the world's biggest single source of electricity—at 27 percent of the total, and cheaper than electricity generated by thermal power plants.

When it is completed in 2018, the Noor solar power plant in the Sahara Desert in Morocco will become the biggest on Earth. Since the Sahara is rarely cloudy there won't be any problem with too little sunshine! Although only part-built, Noor already produces 160 MW of electricity and when it's finished it will make power for 1.1 million people.

A photovoltaic solar farm soaks up the rays, converting them into a supply of electricity.

Energy capture

There are two main kinds of solar power: Photovoltaic (PV) and concentrated solar power (CSP). The photovoltaic effect was first observed by the French physicist Alexandre-Edmond Becquerel in 1839. This effect, which converts light energy to electricity directly, is the basis for modern PV solar power, using solar cells. The way the cells work is pretty

simple. Each one contains two layers of a photovoltaic material, usually silicon. The first layer is "overloaded" with electrons. When photons of light from the Sun hit it, the electrons gain energy, become excited, and move to the second layer, which has the capacity to take on more electrons. This flow of electrons becomes a DC electric current. An inverter converts it to AC current, suitable for running household appliances. The beautiful thing about solar cells is that they have no moving parts to go wrong.

PV solar power can now be generated on any scale. Pocket calculators are powered by a single tiny cell, and people have solar panels on their homes to run air-conditioning and other appliances. Large arrays of solar panels feed electricity into the grid. The biggest such scheme is the Solar Star facility in California, which can generate 579 MW. Innovative schemes such as the Solar Impulse solar-powered aircraft also get their energy from photovoltaic cells.

NASA used solar cells on its spacecraft from the very beginning. For example, Explorer 6, launched in 1959, had four arrays that folded out once in orbit. They provided power for months in space.

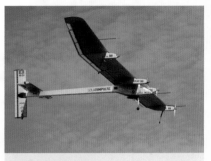

SOLAR IMPULSE

The Solar Impulse aircraft are Swiss-designed single-seat monoplanes powered entirely from photovoltaic cells. The cells power takeoff and cruising flight. On Solar Impulse 2, photovoltaic cells cover the top of its fuselage and 236-foot (72 m) wingspan, powering four electric motors and producing a top speed of 87 mph (140 kph). It is slower at night to conserve energy. In 2015, the aircraft flew nonstop from Japan to Hawaii, the longest-ever journey by a solar-powered plane.

Focused light

Concentrated solar power (CSP) produces electricity indirectly. The idea behind it goes back to ancient Greece when Archimedes supposedly designed a weapon that used mirrors to concentrate sunlight on enemy ships. Professor Giovanni Francia designed and built the first practical CSP plant in Italy in 1968. Parabolic mirrors concentrate the Sun's rays in a solar oven to drive a heat engine. This heats water to create steam, and the steam drives turbines to generate electricity. The largest CSP plant is at Noor in Morocco, and this can generate 392 MW. However, Spain produces more CSP than any other country.

Engineering: the basics

SO WHERE DOES ALL THIS INNOVATION GET US? Let's take a closer look at some primary fields of engineering from a different perspective: Engines, Transportation, Structures, and Materials.

Engines

Steam The first practical engine was powered by steam. The steam engine, invented at the end of the 17th century, and steadily improved throughout the 18th by the likes of Thomas Newcomen and James Watt, is an external combustion engine. That means the fuel is burned outside of the engine's moving parts. The heat from that fuel is used to boil a supply of water, forming a high-pressure stream of steam. That flow of steam is directed into a cylinder, where it pushes on a piston, forcing it to move. The back and forth motion of the piston is then transmitted to machinery, very often converted to rotation to drive wheels. Throughout the 19th century and into the early 20th century, steam engines were the primary source of power for railroad locomotives, tractors, cranes, and buses. However, they were too large and cumbersome for use in smaller vehicles.

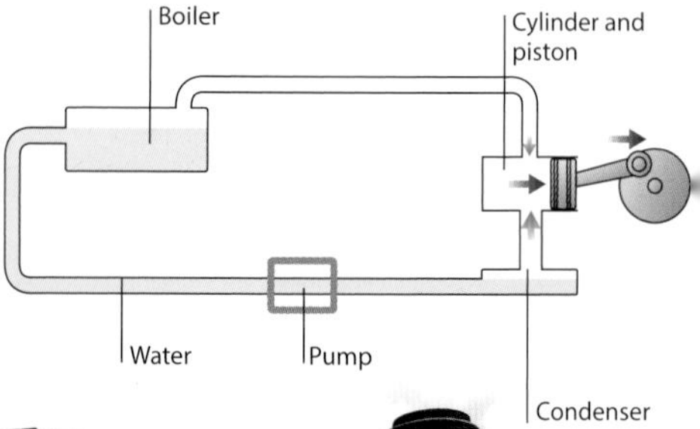

A steam engine in diagram form: A flow of steam from the boiler pushes on a piston creating motion. The steam then condenses, returns to the boiler, and does it all over again.

Boiler | Cylinder and piston

Water | Pump | Condenser

Internal combustion

Internal combustion The first cars, motorbikes, and even the first powered aircraft used internal combustion engines. The engine's invention is credited to Étienne Lenoir in 1858, and by the 1890s it was being used to drive small, personal vehicles—or cars. As its name suggests, the liquid fuel is burned inside the engine's cylinders. It is mixed with air and set alight, creating a supply of fast-expanding gases. These gases push on pistons to create the motion that is then transmitted to the wheels. Most cars use four cylinders, working in the four-stroke cycle (see page 63).

The primary use of the internal combustion engine is to power a car, but they are also used in boats, portable electricity generators, and heavy-duty power tools.

The up-and-down, reciprocating motion of the engine's pistons is converted to a rotational motion

using a crankshaft. This is a zigzagged axle that allows one piston to rise while another falls.

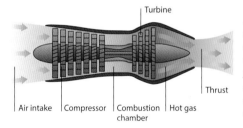

Crankshaft is offset by "throws" | Pistons connect on alternate sides | Pistons move up and down | Crankshaft rotates

Jet The jet engine is a so-called reaction engine, because it creates a thrust force as a reaction to the action of its exhaust gases being pushed out the back. The engine compresses a supply of air and vaporized fuel so it releases a large amount of energy as it burns. That energy is used to spin a turbine, which drives the compressor, and then it goes on to create thrust. The fastest jet aircraft use turbojet engines, but passenger planes are powered by less powerful, but more efficient turbofan engines.

Turbine

Air intake | Compressor | Combustion chamber | Hot gas | Thrust

Fast fighter jets use turbojet engines: A compressor heats and squeezes the air before it mixes with the fuel. The flow of gases produced spins a turbine, which drives the compressor, and then the jet of gas creates thrust.

Rocket All the other engines mentioned here are "air-breathing," which means they need a supply of air (or the oxygen in it) to burn fuels. The rocket engine does not need air. It uses two fuels—a propellant and an oxidant—which burn when mixed. This violent chemical reaction creates thrust, and works in air, in water, or in the vacuum of space. Rocket engines may seem very simple. However, rocket engineers have to design craft that can

withstand the great forces of rocket power and stay under control when traveling at speeds many times the speed of sound.

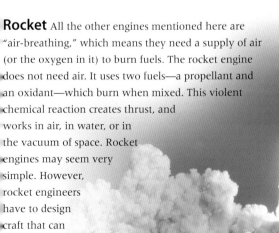

A liquid-fueled rocket engine is relatively simple, with two liquids pumped into a combustion chamber. There they explode, creating a powerful blast of gas.

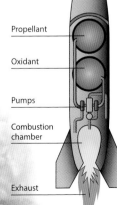

Propellant

Oxidant

Pumps

Combustion chamber

Exhaust

Transportation

Road The invention of the wheel led directly to another—the road. Wheeled vehicles can carry heavier loads than a person or animal, but they can only roll along smooth and relatively flat surfaces. Extensive road networks were built by the Persians and Romans. The most important were paved in stone, but most had rough surfaces simply cleared of rocks.

The modern road was developed by the work of British engineers: In the 1800s, Thomas Telford and John

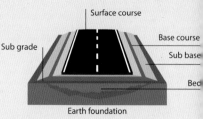

Road vehicles are designed to roll along flat and smooth surfaces—roads in other words. Modern road design is the result of many innovations over centuries.

MacAdam used standard methods for building sturdy foundations topped with compacted small stones. This process became known as macadamization.

In 1901, Edgar Hooley patented a method of bonding the stones with hot tar. This "tar-macadam" or "tarmac" cooled into a flat, hard surface. The stones made the surface slightly rough—ideal for gripping tires. Today, most roads—including runways and race tracks—are surfaced with tarmac.

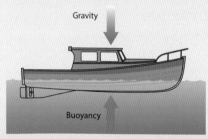

Watercraft float due to the force of the water pushing in the opposite direction to gravity. As long as the total density of the boat is less than the density of water, it will float. Even immense steel vessels are filled mostly with air, and thus weigh less than the equivalent volume of water.

Water Traveling by water was the first mode of long-distance transportation. The simplest rafts as well as the largest tankers and aircraft carriers all rely on the same scientific principle, named for Archimedes who figured it out in the 3rd century BCE. The Archimedes principle states that the force of water pushing on an object (the buoyant force) is the same as the weight of water that object displaces. So the water displaced by a dense object, like a rock, weighs less than the object. This means that the weight of the object outweighs the pushing force of the water—and so the rock sinks. A boat—of any size—will float if it displaces a weight of water that is greater than its own.

Once floating, a boat needs to move—powered by oars, sail, or propeller—and that means pushing its way through the water. The resistance, or drag, of the water will always limit a vessel's speed. The fastest watercraft, such as speedboats, catamarans, and hydrofoils, are designed to use wing-like designs that lift them above the water, and so reduce drag.

Lift

Thrust

Weight

Drag

Air Traveling by air has many advantages. Free from obstacles on the ground, aircraft can move safely at great speeds, and can travel directly to a destination. However, to get into the air, a flying machine needs a lift force that counteracts the pull of gravity, or weight. One way to achieve this is to design a craft that is lighter than air, such as a balloon or airship. However, while these do float upward, they are not easy to propel forward and steer while in the air. Airplanes are heavier-than-air,

and they create lift using a wing. This is a surface with a specific curved shape that cuts through the air. As it does so, the pressure of the air above the wing is made lower than the pressure below. This results in a force that pushes the wing—and the aircraft up. To achieve a lift force that is greater than the weight of an aircraft, it needs to move through the air at high speed. This is achieved by an engine—a propeller or jet—which creates thrust that pushes the aircraft through the air.

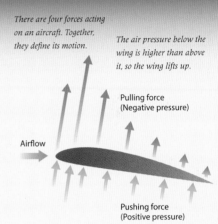

There are four forces acting on an aircraft. Together, they define its motion.

The air pressure below the wing is higher than above it, so the wing lifts up.

Pulling force (Negative pressure)

Airflow

Pushing force (Positive pressure)

Rail Running a vehicle along a railroad track is more efficient that using a road. Trains can carry much greater loads at faster speeds than road transportation. Railroad cars and locomotives run on smooth, metal wheels along metal tracks. Such a system needs to be very heavy to generate enough grip, but once moving, the smooth rails offer very little

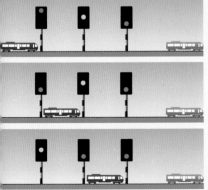

resistance, and trains can carry hundreds of people or very large cargo loads much more efficiently than any other form

Railroads use block signals. A green signal shows the section of track ahead is empty. A red signal shows a train is in the section ahead, and the driver

should stop. An orange light warns of a red signal ahead. Points—literally pointed rails—connect to main rails so trains can switch tracks.

of land transportation. The efficiency has been further increased by replacing steam locomotives with diesel and electric ones. However, there are some big disadvantages to rail transportation. It costs a large amount of money to set up an extensive railroad network with suitable rolling stock, and the trains can only travel where the track takes them.

Structural Engineering

This form of engineering is concerned with the forces at work inside a structure, such as a house, tunnel, or bridge. Engineers ensure that the materials and designs used in these structures are strong enough to hold the weight.

Loads The forces on a structure are known as loads. These include the weight of the structure itself and of any other objects that might use the structure. Structural engineers design a building so the loads are transmitted through structural elements to the ground. Horizontal elements, such as beams or joists, transmit the loads to vertical ones—walls, columns, or posts. These in turn transmit the load to the ground, generally into the foundations, which form a solid base for the structure.

Space The strongest structures would be entirely solid, but that would not be very useful! Instead, the structural elements form a strong frame around the internal

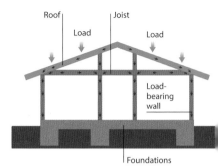

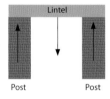

Above: A house is designed to hold the weight of its roof, walls, and floors. All that weight force must be transmitted down through the structure into the ground—or the house falls down!

In a post and lintel system, the lintel pushes down but the posts push up. This system cannot support a wide span because the opposing forces get out of balance, and the lintel snaps.

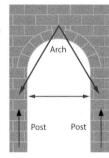

An arch can support wider spans than lintels. The curved structure does not push straight down but to the side. This force is balanced by the upward force of the posts, and the sideways push of the surrounding walls—so the structure remains stable when holding larger loads.

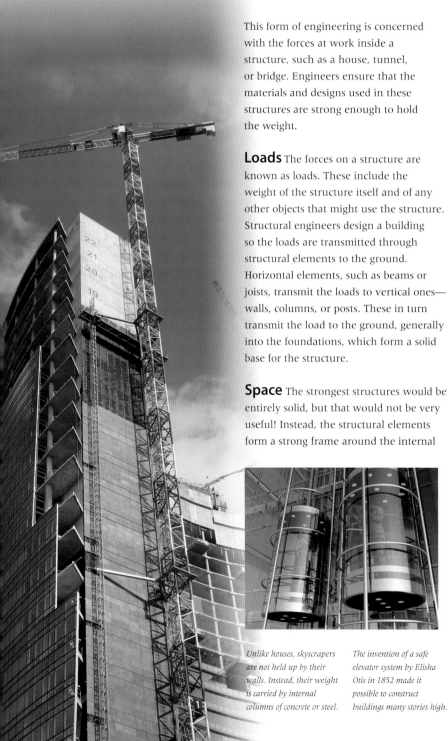

Unlike houses, skyscrapers are not held up by their walls. Instead, their weight is carried by internal columns of concrete or steel.

The invention of a safe elevator system by Elisha Otis in 1852 made it possible to construct buildings many stories high.

space. Narrow spaces, such as doorways, windows, or regular-sized rooms, can be supported using something like a post and lintel system. Here vertical posts hold up horizontal beams or lintels. When a larger, grander space has to be spanned, a post and lintel will not be strong enough. The horizontal elements cannot transmit all the load to the posts so begin to bend—and break. Larger spaces are spanned with arches. Their curved structures are better at directing load forces safely into the vertical elements of the building.

As a building gets taller, its weight increases and the walls that bear that load must get thicker. Above about 12 stories, this becomes impractical because walls would need to be several feet thick. So to go higher, structural engineers design buildings that are supported by massive internal columns. The floors stick out from the column and the walls are hung around the outside to create internal spaces, but not to carry any of the weight of the building itself.

Spans The most challenging structures are bridges that span large gaps. They can have only a limited number of load-bearing supports, but must be strong enough to hold their own weight—and the weight of whatever is crossing them. Small-span bridges use simple beams or arches made of wood or stone. However, to cross larger spaces, these materials are too weak to hold their own weight. Instead, bridge

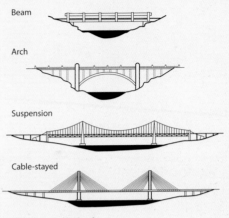

Beam

Arch

Suspension

Cable-stayed

Four designs of bridge, from top to bottom capable of crossing ever larger spans.

builders turn to steel, which can bend and flex without breaking. Large-span bridges use a number of designs that transmit the weight of the central deck (or roadway) out to the ends of the bridge that are connected to the ground. A cantilever bridge works like a pair of teeter-totters, that balance the weight of the central deck. A suspension bridge hangs the deck from sturdy cables that extend from both ends. Cable-stayed designs do something similar, but hold the deck by hanging it from tall, upright supports.

Above: Construction workers building the Forth Bridge in Scotland in the 1880s show how the railroad bridge's two sides would balance the central span between them.

Below: The Golden Gate Bridge across the entrance to San Francisco Harbor is one of the most famous suspension bridges in the world. It was opened in 1932, and was the largest in the world until 1964 when the Verrazano–Narrows Bridge in New York opened.

Materials

Stone The oldest structures that still stand are made of stone, which is testament to the great strength of this material. In general, stone is a hard material. It withstands huge compression forces and so will not break when bearing large weights. However, stone is weak when under tension or if required to bend. It will crack when needed to support a wide span, and therefore this limits what can be built from stone.

Above: Incan architecture in Cusco, Peru. Quite how Incan stonemasons cut stones to fit so exactly remains a mystery.

Wood Wood is probably the oldest building material. Being organic it eventually decays and so few wooden buildings remain standing for more than a few centuries. However today, wooden beams, joists, and posts are integral elements in most modern houses. It has the advantage of being inexpensive, and is tough enough to be strong under the kind of compression and tension forces found in a house structure—it will bend a little but does not crack. However, wood is less suitable in cold, damp climates, where it is prone to rotting.

A traditional Japanese house has a wooden frame and no internal load-bearing walls—inner walls are made of paper.

Mud Simply a mixture of earth and water, mud is a very inexpensive building material in dry climates, and surprisingly versatile. It is applied while wet and built up in layers to create walls and floors. Once dry it is strong enough to hold a wooden roof. Some of the earliest structures were made of mud, reinforced with woven sticks inside—a system called wattle and daub. The disadvantage of mud structures is that they are damaged by rain and must be maintained frequently.

Below: The Great Mosque at Djenné, in Mali, West Africa, is made of mud and is 800 years old.

Concrete Mimicking the structure of natural rock, concrete is made of sand or small stones that are glued together by a chemical cement. Several types of cement are used, which are designed to work in different conditions, but the results are largely the same. Concrete is very strong under compression and so can be used to bear enormous loads, such as in dams and skyscrapers. Concrete is wet when mixed and is poured into molds (called forms) where it sets solid (goes off). Therefore, it can be molded into all kinds of shapes that have the durability of stone.

A table and chairs made from set concrete. Concrete is a cheap and durable material.

Plaster In construction this material is used to create a smooth finish on walls and ceilings, but it can also be molded to create intricate details and features. Plaster is generally made from powdered gypsum or lime. These naturally occurring chemicals are heated to drive out any water from their crystal structure. When mixed with water, the crystals reform into their original structures. As they do that, the wet plaster can be spread on surfaces or molded before it sets solid.

A ceiling rose made from molded plaster.

Brick buildings are strong under compression and so are good for making long-lasting walls. However, brick buildings cannot be higher than about 12 stories before the wall thickness becomes impractical.

Brick This is the material of choice for masonry structures, which are built from blocks that are layered on top of each other and cemented together to make walls and other structural elements. (Other masonry buildings might use stones cut into blocks.) Modern bricks are made from clay, which is added to molds when wet and baked solid. The high temperatures fuse the clay in a similar way to firing pottery. Bricks have been used for at least 7,000 years and are best suited for small buildings in cold and damp climates.

Bronze A mixture, or alloy, of copper and tin, bronze was the first metal in widespread use. The Bronze Age began in the Eastern Mediterranean around 5,000 years ago. Its constituent metals are relatively unreactive and therefore easy to purify from ores: The temperature required to smelt them can be produced by a large wood fire. Bronze is easy to work, that is to mold and shape, but is stronger than pure copper or tin by themselves. However, bronze objects need to be relatively thick, because they become brittle and snap if made too thin.

A bronze helmet may be weak compared to other metals, but its purpose was to defend against weapons also made from bronze.

As well as being lightweight, aluminum is also resistant to corrosion (unlike steel).

Iron This metal is the second most common metallic element found in Earth's rocks. It requires a higher temperature to purify than copper, but by around 3,000 years ago, charcoal furnaces were able to reach high enough temperatures to smelt it. The purest form of iron is known as wrought iron. A less pure, but less expensive form, is called cast iron. Cast iron is molded into objects while molten. When cooled it forms a strong, heavy object. However, it is brittle and cannot be hammered into shape or worked easily. Wrought iron is easier to work but is also brittle and cracks easily when bent. Cast iron is still used to make pans and similar items, but wrought iron has been superseded by steel.

The Eiffel Tower is made from 18,038 pieces of wrought iron connected by 2.5 million rivets.

Steel This is an alloy of iron and carbon, and generally small amounts of other metals. The addition of carbon crystals (between 1 and 2 percent) makes the metal much stronger than pure iron. Steel was difficult to make until an industrial process was developed in the 1850s. Today, it is the cheapest and most widespread metal used in construction. It is hundreds of times stronger than concrete when under tension.

Aluminum The most common metal in Earth's crust, aluminum is highly reactive and therefore difficult to purify from ores. By the 1870s, an electrochemical process was developed to make aluminum production economical. Aluminum is not as strong as steel but is considerably less dense and so is used to make tough but lightweight objects.

Steel arches are used to support this bridge. The steel must be maintained to prevent rusting (corrosion). A simple way of doing this is to paint the metal to protect it from the elements.

The body of this motorcycle is made from carbon fiber. This makes it strong but lightweight, so the bike is more fuel efficient.

Carbon fiber

This modern material has a very high strength to weight ratio, making it a useful lightweight alternative to metal components. A carbon fiber is a long strand of carbon, and these are woven or glued together. Carbon fiber is a common component of composite materials, where metals, plastics, and other materials are used together. Composite materials can be engineered to have very specific properties to suit a wide range of applications.

A strip thermometer uses a smart material. The color of the material changes with temperature. When pressed on the skin it indicates body temperature.

Plastic

The term "plastic" refers to a material property: A substance that can change its shape without being damaged. Plastic is the term used for many compounds that have this property made from the chemicals in petroleum. Plastics are polymers: In other words, they are made from long, chained molecules. These long molecules can be arranged in many ways to make strong, lightweight materials in just about any shape. Some plastics can be heated and reshaped many times. In others, known as thermosets, the polymers lock in place and cannot be reshaped.

The components in this car engine are not made from metal but from thermoset plastics that stay strong at high temperatures.

Smart materials

This catchall term refers to a wide range of materials that are able to respond to their surroundings and change their properties. The most familiar smart material is quartz. Quartz crystals produce electric current when squeezed, and vibrate when electrified. Other smart materials change their properties with temperature or magnetism. An exciting new property is shape memory. When conditions reach a critical state, such as a certain temperature, the material changes into another pre-engineered shape.

A digital watch keeps time using the rhythmic vibrations of an electrified quartz crystal—an early use of a smart material.

IMPONDERABLES

EVERY ENGINEER IS BUSY BUILDING THE FUTURE, IMPROVING OLD DESIGNS, AND MAKING THINGS BIGGER OR BETTER. What they will build next is limited only by our imagination. Let's take a look at some possible innovations.

Is it possible to engineer a new Earth?

A terraformed planet might be a future home in space.

Evidence of Earth-like planets is beginning to emerge in distant star systems, but we have no earthly means of visiting them. It is conceivable that we could soon send humans to Mars, however. Would it be possible to engineer a new Earth there? The process is dubbed terraforming and would be a gargantuan task. The first problem to solve is Mars's diffuse atmosphere, barely one percent as thick as Earth's. Redirecting nitrogen-rich comets to Mars would be one way of thickening the atmosphere. Next we need oxygen. On Earth this is made by plants taking carbon dioxide from the air. The Martian atmosphere is mostly carbon dioxide, so the planet could be seeded with plant-like bacteria, that would cling to life in rock crevices at first, slowly producing oxygen and adding organic material to the dusty soil. However, Earth's magnetic field protects us from dangerous solar radiation. Mars lacks this life-saving ability, and engineers have yet to propose a way of giving the red planet a magnetic field.

Will the Hyperloop happen?

In 2012, the billionaire computer scientist and all-round futurologist Elon Musk proposed a form of transportation called the Hyperloop. The concept envisages passengers seated in pods, hurtling between cities inside a network of tubes. The tubes would hold a partial vacuum, and a fan at the front of the pod would suck in air and then pump it out through the base. This would create a cushion of air under the pod allowing it to glide along like an air-hockey puck. With little air resistance, the Hyperloop could reach supersonic speeds. Musk has asked several companies to compete for the right to build the Hyperloop. Some say the first Hyperloops will be in service by the mid-2020s. Such a system could reduce the need for polluting short-haul flights. But Musk's idea is not without its problems. A journey through the Hyperloop could be distressing with the loud noises, accelerations, and debilitating vibrations. However, they said the same thing about train travel.

Elon Musk's gliding Hyperloop is said to be the fifth form of transportation, adding to wheels, wings, rails, and boats.

How high can we build?

There are no structural constraints to building a tower taller than Mount Everest. But the base would have to be 1,580 square miles (4,100 square km), which is a lot of real estate for a developer to secure. Taking this further, the tallest skyscrapers are about 15 percent the density of Mount Everest, and so could be nearly seven times taller than the mountain and still weigh the same. Would we ever build such a structure? If it makes financial sense, then we will. The Jeddah Tower in Jeddah, Saudi Arabia, will be 3,281 feet (1,000 m) when completed in 2020. The initial idea was to go to a mile tall. That height was reasonable, but its price tag was too much of a tall order.

The real limit to a building's height is money, not engineering.

Will we get an elevator to space?

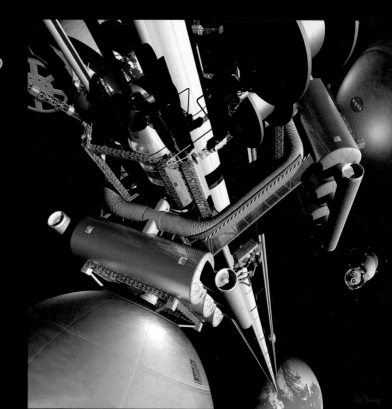

In 1895, Konstantin Tsiolkovsky, a visionary engineer, proposed a space elevator. This is a column that rises from Earth's surface into space. Later analysis showed that such a structure would need to be 43,500 miles (70,000 km) tall. An elevator in the column would carry travelers to a space station about 22,250 miles (35,800 km) above the surface. At this height, objects orbit at the same speed as Earth's rotation and so always stay above the same spot. To hold this geostationary platform steady, a counterweight would be placed much further out in space. This would mean that the tall column was not under compression like a regular building, but was held in place by being pulled tight by the counterweight. But there are no available materials that could be used to build such an elevator. If any could be developed—carbon nanotubes are one possibility—then an elevator would make space travel very inexpensive. While energy is needed to haul loads up the elevator, they fall under

Will robots ever become like humans?

An android is a robot that emulates a human's body and abilities. It moves like us and in future could think like us—or appear to think like us, which is much the same thing. There are two engineering hurdles to making androids match humans, perhaps outmatch them. The first is one of power: The mechanics of an android body and the motors that move every joint are much heavier than the biological version. A robot as agile and flexible as us would also be immensely heavy. One solution would be to grow robot bodies, or the mechanical parts, from biological material. Secondly, our bodies are controlled by our brains (mostly), and an android would be controlled by a computer. There is no evidence that the brain works like a computer. Even if a computer could match the processing power of a brain and be made small enough, how do we program it? The result is androids would fall into the "uncanny valley," a term for humans' negative reactions

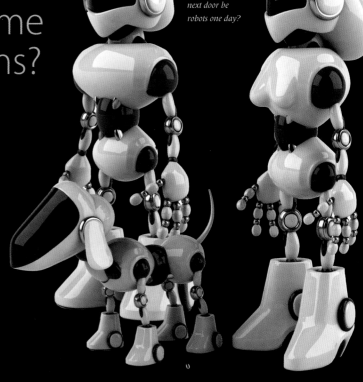

Will the family next door be robots one day?

What is the cleanest fuel system?

Burning a fossil fuel always creates pollution—chemicals that damage health and the environment. Coal is the dirtiest, while pure methane is the cleanest, although even that releases carbon dioxide, which will be the most damaging pollutant of all time if it is allowed to transform Earth's climate. The cleanest non-fossil fuel is hydrogen. When you burn it, the residue is nothing but water vapor. However, pure hydrogen has to be refined from other chemicals. The very act of making it requires a lot of energy (perhaps from fossil fuels), and so in the end hydrogen will not be

Do we need a new type of computer?

A classical digital computer uses a series of switches, or relays, which turn on and off to direct current around a complex circuit. Each switch is made of a semiconductor sandwich, which forms a tiny gap in a circuit. That gap can be made to either block current or let it pass. The size of that gap has been shrinking as engineers can pack more circuitry into devices. However, the gap cannot get much smaller before it offers no barrier to electricity. There are alternatives. Photonic circuits make use of light beams not electric currents, while the most exciting possibility is to use individual atoms as the switches. Because of quantum behavior, these switches could be on and off at the same time. This is the basis of quantum computing. If atoms could be connected together in groups they would have vast processing power. Two classical switches store 2 bits of information, while two quantum switches store 4 bits (2^2), and a 64-quantum bit system can handle not 64 bits but 18,446,744,073,709,551,616!

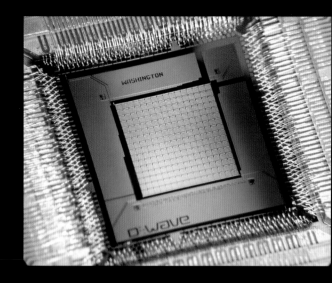

The D-Wave quantum processor, sponsored by NASA and Google, is said to be the first working quantum computer.

A sturdy balloon that is inflated in space may be the future of spacecraft design.

Will future spacecraft be balloons?

In 2016, a new module was added to the International Space Station (ISS). The BEAM (Bigelow Expandable Activity Module) was just 8 feet (2.4 m) wide when it was launched into space. However, it was then inflated with air to make it five times the size, providing another room for the ISS crew. The expandable walls are made from several layers of kevlar, the material used in bulletproof vests. The BEAM will be tested in space for two years. It is hoped that the kevlar walls will provide better protection from radiation than the traditional aluminum modules, and will prove tough enough to withstand impacts from dust and space debris hurtling around the ISS's orbit.

IMPONDERABLES

Will spaceplanes change transportation?

Spaceplanes might make it possible to take a day trip to the other side of the world.

Aircraft must push themselves through the air and are slowed by air resistance. If passenger craft could get out of the atmosphere, the craft would reach enormous speeds in a suborbital path that would allow it to cross oceans and continents in minutes, not hours. A few companies are developing such craft, and engines, that could perform this feat. One is the SABRE engine from the UK engineers Reaction Engines. This engine has a three-in-one design using a jet engine, a ramjet, and rocket. The jet's fuel is chilled so it produces more thrust when burned. The ramjet would join in once the spacecraft had reached a high speed through the air. The high-speed air rushes into the ramjet and fuel is burned without the need for a compressor. The rocket does not need air and would push the aircraft into space.

What will graphene do for us?

Graphene is a manufactured form of pure carbon made from graphite. Graphite is better known as pencil lead. It is made from loosely bonded layers each made up of an interconnected lattice of carbon atoms arranged in hexagons. The layers slide over each other, giving graphite its soft and slippery properties. Graphene is a single layer of graphite. Research has begun into what graphene can do: Rolling the sheets into tubes could result in superstrong construction materials. It might be used to make tiny machines that can work inside the body, or be used in the next generation of batteries, computer chips, displays, or solar panels. Graphene is heralded as the material that will make the future. Time will tell.

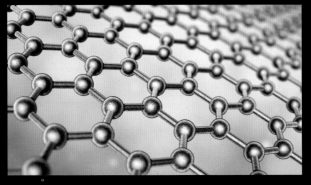

Graphene's hexagonal structure makes it strong enough to form sheets that are only one atom thick.

Will we run out of raw materials?

It is possible that Earth's resources will run out, especially the rare metals used in our high-tech electronics. Asteroids are the rocky remnants of the early Solar System. In 2001, NEAR Shoemaker, a tiny spacecraft, made the first and only soft landing on an asteroid. Plans are already afoot to send larger landers to asteroids—ones that are mostly made of metals. The gravity of an asteroid is a fraction of that of Earth, which means it could be possible to load up a return craft with metal-rich rocks, containing all kinds of valuable materials that are in short supply on Earth, and launch it into space to return to Earth.

Can screens replace paper?

Paper is lightweight, foldable, and strong. The only problem is once you've written or printed on it, you need another piece. Screen displays can be refreshed, but they are heavy, rigid, and require power. If a screen could be made flimsy like paper then perhaps the wood-based medium, one of humanity's greatest inventions, would become a thing of the past. Step forward the OLED, or organic light-emitting diode. This display system uses an ultrathin layer to produce patterns of colored lights. The layers are so thin that they can be placed on flexible plastic. OLED displays are already used for curved screens. One could imagine touch-sensitive OLED "paper" powered by photovoltaics.

Flexible displays might one day replace paper.

Can engineering solve climate change?

Carbon dioxide added to the atmosphere by human activities prevents Earth from radiating enough heat away. This extra heat will most likely result in a worsening of weather extremes. In the light of this, cutting carbon emissions is just the first step. Climate engineers are developing ways of reversing the changes. The first methods cut the sunlight hitting Earth. Vast mirrors could unfold in space to reflect the light away, or aircraft could pump fine powders into the sky to block out the light. Another goal is to reduce the carbon dioxide in the air. The gas could be extracted using chemicals and pumped underground. Or pulling up nutrients from deep waters and adding iron-rich fertilizer would boost oceanic algae, which could convert the carbon dioxide into biomass, some of which would sink to the seabed. Climate engineering could be the largest project ever undertaken by humans.

An artist's impression of a futuristic mining colony on an asteroid.

The Great Engineers

ENGINEERING IS AS OLD AS CIVILIZATION ITSELF BUT THE NAMES OF THE INNOVATORS WHO BUILT the first wheels, levers, and pulleys are lost to history. In more recent times, the achievements of the greatest engineers have ensured they are remembered down the ages. However, engineering is seldom the work of one mind alone. The design, testing, and construction is all carried out by a team of people, who add ideas and check the work of others. Unlike science, where great thinkers are happy to be proven wrong by a future discovery, engineers need to be sure that their creations are fit for purpose—and that takes teamwork.

Imhotep

Born	27th century BCE
Birthplace	Memphis, Egypt
Died	27th century BCE
Importance	Architect of first Egyptian pyramid

Imhotep was the first architect we know of by name. He was one of Pharaoh Djoser's most trusted advisers and the designer of his ruler's pyramid tomb. Imhotep's influence extended beyond his lifetime—he was venerated as the patron of scribes, and was also described as the son of Ptah, chief god of Memphis. Later, Imhotep was even raised to godhood himself, worshiped in Memphis, where he was served by his own priesthood and was said to act as an intermediary between gods and humans. Imhotep was also reckoned to be a skilled physician who extracted medicine from plants and treated diseases such as appendicitis and arthritis.

Archimedes

Born	c.287 BCE
Birthplace	Syracuse, Sicily
Died	c.212 BCE
Importance	Inventor and mathematician

For many, Archimedes' greatest claim to fame may be his cry of "Eureka!", supposedly made as he stepped into his bath and in a flash of inspiration and displaced water discovered the law of hydrostatics. He is credited with inventing the compound pulley and the hydraulic screw for raising water, a device still used today. During the Roman conquest of Sicily in 214 BCE Archimedes is said to have set light to the invaders' ships by focusing the rays of the Sun on them. When Syracuse eventually fell, Archimedes was killed by a Roman soldier, who ignored Archimedes' request not to be disturbed in his calculations.

Vitruvius

Born	1st century BCE
Birthplace	Fundia, Italy
Died	1st century BCE
Importance	Influential architect

Roman architect and engineer Vitruvius is most celebrated for his treatise *De architectura* (*On Architecture*), which he dedicated to an unnamed emperor, most likely Augustus. It is almost certain that Vitruvius would have been born into a wealthy family. He served with the 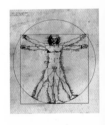 engineering corps in Julius Caesar's army. In 46 BCE, he was in Zama in North Africa; the Romans were building a city there and it is likely that Vitruvius was involved. After Caesar was assassinated in 44 BCE, Vitruvius joined the army of Octavian (later to become the Emperor Augustus) as a military engineer. By 33 BCE, he was engaged in building aqueducts.

Heron

Born	c.10 CE
Birthplace	Alexandria, Egypt
Died	c.75 CE
Importance	Inventor of first steam-powered device

Heron (or Hero) of Alexandria was a Greek mathematician and engineer with an evidently playful ingenuity and inventiveness. In his book *Pneumatica*, he describes over a hundred marvellous mechanical devices, including singing birds, dancing puppets, a water-powered organ, coin-operated machines, and a fire engine, in addition to his famous steam-powered aeolipile. Probably Heron built these not for simple amusement but as teaching aids to demonstrate theories to his students. His writings are a valuable source of information on the engineering and mathematical knowledge of Babylon, ancient Greece and ancient Egypt.

Cornelis Drebbel

Born	1572
Birthplace	Alkmaar, Netherlands
Died	November 7, 1633
Importance	Inventor of the submarine

Cornelis Drebbel (or Cornelius van Drebbel), inventor of the submarine, was apprenticed to an engraver but his interests turned to alchemy and engineering. In 1598, he was granted a patent for a supposedly "perpetual motion" machine, which he claimed utilized changes in the air and tides to power a clock. Whether it worked like that or not, it certainly made a name for him. Drebbel was a skilled lens grinder and is said 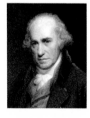 to have made the first compound microscope. Around 1604, he was invited to England by King James I, who was keen to surround himself with learned courtiers. After James's death Drebbel was employed making "secret weapons" for King Charles I.

Zhang Heng

Born	78 CE
Birthplace	Nanyang, China
Died	139 CE
Importance	Inventor of the seismograph

Astronomer, mathematician, engineer, painter, and poet, Zhang Heng is also remembered as the inventor of the seismograph. He was educated in the philosophy of Confucianism. For ten years he trained as a writer, publishing a number of literary works and winning considerable fame in the process. Zhang was thirty years old before turning from literature to science, in particular astronomy. Around 116 CE he was appointed an official at the Emperor's court, eventually becoming chief astrologer and minister. The Chinese believed that civil life was reflected and controlled by the heavens, so the role was an important one.

James Watt

Born	January 19, 1736
Birthplace	Edinburgh, Scotland
Died	August 25, 1819
Importance	Inventor of practical steam engine

James Watt may have been a brilliant engineer but by his own account he was a terrible businessman. In his own words: "I would rather face a loaded cannon than settle an account or make a bargain." Watt's father was a prosperous shipbuilder, but a shipwreck cost the family dear and Watt had to learn a trade. He became a maker of scientific instruments. At Glasgow University he was given access to a Newcomen steam engine, drastically improving its design. The engineering firm he set up with his partner Matthew Boulton was the most important in the country. Watt retired a wealthy man and devoted his last years to research.

George Stephenson

Born	June 9, 1781
Birthplace	near Newcastle-upon-Tyne, England
Died	August 12, 1848
Importance	Influential railroad engineer

George Stephenson, the son of a Newcomen engine operator, had no formal schooling and learned to read and write at night while working at a coal mine. As a married man he bolstered his income by fixing clocks and repairing shoes. But it was with steam engines that his true talent lay. In 1813, he built the Blucher, a steam-powered locomotive that could haul 30 tons of coal at 3.7 mph (6 kph). Stephenson built several more locomotives over the next few years and constructed the Stockton to Darlington line in 1825, the first railroad to connect two towns. He worked on several other railroad projects in England, Spain, and Belgium.

John Roebling

Born	June 12, 1806
Birthplace	Mühlhausen, Prussia (now Germany)
Died	July 22, 1869
Importance	Designer of Brooklyn Bridge

John Augustus Roebling emigrated to the United States from Germany at the age of 25. While working for the Pennsylvania Canal system, he experimented with making a cable of twisted wires to replace the hemp ropes used to haul the canal boats. The success of his wire rope led to him launching a business that made his fortune. His cables were used in suspension bridges, most famously the Brooklyn Bridge. His most famous achievement was also what killed him. He crushed his toes while surveying the site and poured water on the wound. Unfortunately, he used unsterilized well water and three weeks later he died of tetanus.

Isambard Kingdom Brunel

Born	April 9, 1806
Birthplace	Portsmouth, England
Died	September 15, 1859
Importance	Designed bridges and oceangoing steamships

Brunel was without doubt one of the greatest and most ambitious engineers of the nineteenth century, turning his talents to ships, tunnels, railroads, and bridges. He was the only son of French engineer Marc Brunel. In 1825, not yet 20, Brunel was appointed resident engineer on the Thames Tunnel project. In 1831, he won the competition to design Bristol's Clifton Suspension Bridge but although work began in 1834 it was not completed until after Brunel's death. Among his other achievements was the Great Western Railway linking Bristol to London.

Gustave Eiffel

Born	December 15, 1832
Birthplace	Dijon, France
Died	December 28, 1923
Importance	Builder of Eiffel Tower

Gustave Eiffel is not only known for the Paris tower that bears his name but for several other projects, including the Statue of Liberty and the 531-foot (162-m) steel arch of the Garabit Viaduct in France. Eiffel specialized in metal construction. He designed the impressive arched Gallery of Machines for the Paris Exhibition of 1867. His career took an unfortunate turn when he was involved in a failed attempt to construct a canal across Panama in 1887. Although not directly responsible, Eiffel was charged with misuse of funds and narrowly escaped prison. He later turned his attention to aerodynamics, a field in which he made several important discoveries.

Karl Benz

Born	November 25, 1844
Birthplace	Karlsruhe, Baden, Germany
Died	April 4, 1929
Importance	inventor of first automobile

Karl Benz's father, a locomotive driver, died in an accident when Benz was two. He enrolled at the University of Karlsruhe at age 15 and graduated in 1864 with a degree in mechanical engineering. After a succession of jobs, Benz married in 1872 and concentrated on the development of new and better engines, inventing battery-powered ignition systems, spark plugs, and the clutch, among other things, all of which contributed to the success of his Motorwagen car in 1885. The first long distance automobile journey was made by Benz's wife, who borrowed the Motorwagen to visit her mother.

Alexander Graham Bell

Born	March 3, 1847
Birthplace	Edinburgh, Scotland
Died	August 2, 1922
Importance	Inventor of the telephone

Alexander Graham Bell was fascinated by the transmission of sound. His mother was almost totally deaf and his father was a teacher of deaf children, a career that Bell also embraced. He never completed any formal education as he left school at age 15, and although he gained entrance to University College, London, the Bell family moved to Canada before he completed his studies. The patent Bell was granted for the invention of the telephone made him an exceedingly wealthy man. He was keen to foster the spread of scientific knowledge—he was a founder of the National Geographic Society and a supporter of the journal *Science*.

Thomas Edison

Born	February 11, 1847
Birthplace	Ohio, USA
Died	October 18, 1931
Importance	Prolific developer of electrical devices

Edison was one of the most prolific inventors who ever lived, with more than 1,000 patents to his name. Partial deafness made school difficult for him, so in 1859 he left school and began work as a trainboy on the railroad between Detroit and Port Huron, later becoming an apprentice telegrapher. He moved to New York in 1869 where he became involved in the telegraph industry. In 1876, he built a laboratory at Menlo Park, New Jersey. This was to be where many of his most important inventions were developed, including the phonograph and the electric light.

Nikola Tesla

Born	July 10, 1856
Birthplace	Smiljan, now Croatia
Died	January 7, 1943
Importance	Prolific electrical engineer

Tesla was an engineer and inventor of real genius, who did much to shape the modern world. He studied electrical engineering at the Polytechnic Institute at Graz, Austria, where he delighted in arguing with his professors. Unfortunately, Tesla became addicted to gambling, lost all his tuition money, and dropped out. In 1884, Tesla sailed for New York City to work with Thomas Edison. The two soon parted company, and Tesla went to work for Edison's rival George Westinghouse, who used his AC system for transmitting electricity. Tesla went on to develop many more electrical and radio devices. However, he died in poverty in a New York hotel.

Henry Ford

Born	July 30, 1863
Birthplace	Dearborn, Michigan, USA
Died	April 7, 1947
Importance	Developer of mass production of cars

Henry Ford was one of eight children, born on the family farm to the west of Detroit. At the age of 16 he walked to Detroit to find work and made his first acquaintance with the internal combustion engine. In 1893, Ford became chief engineer at the Detroit Edison Company electricity plant. After years of tinkering with his designs in his spare time, the Ford Motor Company put its first car on the market in 1903. The Model T was launched in 1908 and by 1918 half of all cars in America were Fords. Ford was a controversial figure. He was a sworn enemy of labor unions, was an avowed anti-Semite, and was an admirer of Adolf Hitler.

Robert H. Goddard

Born	October 5, 1882
Birthplace	Worcester, Massachusetts, USA
Died	August 10, 1945
Importance	Inventor of liquid-fueled rocket

As a teenager, Robert H. Goddard climbed into a cherry tree and daydreamed about the possibilities of sending a rocket to Mars. He later began a lifetime of experiments with space rocketry and was given a grant by the Smithsonian Institution to further his research. *The New York Times* rubbished Goddard's ideas about space flight, yet published a belated

apology in 1969 when Apollo 11 flew to the Moon powered by a huge liquid-fueled rocket. On March 16, 1961, Esther Goddard was present for the opening of NASA's Goddard Space Flight Center, 35 years to the day after her husband launched the first liquid-fueled rocket.

Guglielmo Marconi

Born	April 25, 1874
Birthplace	Bologna, Italy
Died	July 20, 1937
Importance	Inventor of radio communication

The son of an Italian father and an Irish mother, Marconi studied physics at the Livorno Technical Institute. He founded the Marconi Wireless Telegraph Company in London in 1899. In December 1901, he sent the first wireless message across the Atlantic, beginning a new age in communication. He was awarded the Nobel Prize in Physics in 1909 for his work. Although Marconi himself wasn't very interested in television, his company joined with EMI to develop the system that was adopted in 1936 by the British Broadcasting Company for its launch of the world's first high-definition television service.

John Logie Baird

Born	August 13, 1888
Birthplace	Helensburgh, Scotland
Died	June 14, 1946
Importance	Inventor of the television

John Logie Baird demonstrated early signs of his technical expertise when he set up a telephone exchange connecting his bedroom to those of his friends across the street. As ill health made him unfit to fight, he spent World War I as an electrical engineer. After the war he moved to England and carried on his experiments in television, eventually giving the first demonstration in London in 1926. By the 1930s, Baird's mechanical system had been superseded by an electronic one. Undaunted, Baird experimented with color television and even 3-D transmissions.